S0-AUM-572

Clinton,

December 1983.

BIMETALLIC CATALYSTS

BIMETALLIC CATALYSTS

Discoveries, Concepts, and Applications

JOHN H. SINFELT
Corporate Research Science Laboratories
Exxon Research and Engineering Company

An Exxon Monograph

JOHN WILEY & SONS
New York • Chichester • Brisbane • Toronto • Singapore

Copyright © 1983 by John Wiley & Sons, Inc.

All rights reserved. Published simultaneously in Canada.

Reproduction or translation of any part of this work beyond that permitted by Section 107 or 108 of the 1976 United States Copyright Act without the permission of the copyright owner is unlawful. Requests for permission or further information should be addressed to the Permissions Department, John Wiley & Sons, Inc.

Library of Congress Cataloging in Publication Data:

Sinfelt, John H. (John Henry), 1931–
Bimetallic catalysts.
Includes index.
1. Metal catalysts. 2. Catalysis. I. Title.
QD505.S59 1983 541.3'95 83-6738
ISBN 0-471-88321-2

Printed in the United States of America

10 9 8 7 6 5 4 3 2 1

FOREWORD

Exxon's scientists and engineers have a long tradition of significant inquiry and achievement. Their findings appear in the leading journals of the world, but until now no means existed to present their work in the context in which it was conducted. This and future Exxon Monographs are meant to provide a thorough look at how we address research problems and the results of this research.

Bimetallic Catalysts: Discoveries, Concepts, and Applications is an appropriate vehicle for launching the Exxon Monograph program. Its subject has major industrial significance: bimetallic catalysts are the heart of today's boosters required for the manufacture of unleaded gasolines. The author, John H. Sinfelt, is a leading researcher in bimetallic cluster catalysis, and in 1979 he received the United States government's highest scientific award, the National Science Medal, for his contribution to catalysis research.

The Exxon Monograph program reflects the efforts of many people at Exxon Corporation and Exxon Research and Engineering Company, including its first editor, Ben H. Weil, now retired. In addition, to ensure that each monograph is a work of scholarship as well as a record of industrial research, both the outline and the completed manuscript undergo extensive internal and external peer review before publication.

We hope that by sharing our expertise in solving important scientific, engineering, and technical problems in this way, Exxon will help advance the knowledge necessary for future discovery and innovation.

E. E. David, Jr.
President, Exxon Research and
Engineering Company

PREFACE

This monograph is an account of research on bimetallic catalysts conducted in the laboratories of the Exxon Research and Engineering Company from the early 1960s to the present. The research began in the old Process Research Division of the company. Subsequently, it was transferred to the Central Basic Research Laboratories, the forerunner of the Corporate Research Laboratories organized in 1968. Most of the research discussed here was conducted in the Corporate Research Laboratories.

Although bimetallic catalysts did not represent a totally new area of research in the early 1960s, my research emphasized entirely new aspects of this subject. Earlier work on metal alloy catalysts was dominated by efforts to relate the catalytic activity of a metal to its electron band structure. Very little attention had been given to other aspects of metal alloy catalysts, such as the possibility of influencing the selectivity of chemical transformations on metal surfaces and of preparing metal alloys in a highly dispersed state. These aspects were the basis for my work on bimetallic catalyst systems.

During the research, I introduced the term "bimetallic clusters" to refer to highly dispersed bimetallic entities present on the surface of a carrier such as silica or alumina. The term "highly dispersed alloys" did not seem appropriate as a general name for such entities, since systems of interest include combinations of metals that are immiscible in the bulk and therefore do not form alloys in the usual sense. Much of the material included in the monograph is concerned with the validation and elucidation of the bimetallic cluster concept, beginning with the use of chemical probes such as chemisorption and reaction kinetics in the early stages of the work and extending to the use of physical probes such as Mössbauer effect spectroscopy and extended X-ray absorption fine structure (EXAFS) in the later stages.

Research on bimetallic catalysts has had a major impact in the reforming of petroleum naphtha fractions to produce high octane number components for gasolines. During the 1970s, bimetallic catalysts largely replaced traditional platinum catalysts in reforming. This development has been an important factor in the emergence of "low-lead" and "lead-free" gasolines.

My work on bimetallic catalysts has been aided greatly by a number of collaborators, including J. L. Carter, A. E. Barnett, G. W. Dembinski, W. F. Taylor, D. J. C. Yates, G. H. Via, J. A. Cusumano, R. L. Garten, Y. L. Lam, C. R. Helms, F. W. Lytle, G. B. McVicker, E. B. Prestridge, W. Weissman, W. S. Kmak, S. C. Fung, and G. Meitzner. It is a pleasure to acknowledge their valuable contributions to the work. I also wish to thank various persons in the management of the Exxon Research and Engineering Company, especially Dr. P. J. Lucchesi and Dr. A. Schriesheim, for their encouragement in the initial stages of research and for their continuing support throughout.

For permission to reproduce, in whole or in part, certain figures and tables, I am grateful to the following publishers: Academic Press, American Institute of Physics, North-Holland Publishing Company, Elsevier Scientific Publishing Company, and the American Chemical Society.

JOHN H. SINFELT

Linden, New Jersey
July 1983

ABOUT THE AUTHOR

John H. Sinfelt is a Senior Scientific Advisor in the Corporate Research Science Laboratories of Exxon Research and Engineering Company, Linden, New Jersey. He has been active in heterogeneous catalysis research during his entire career at Exxon, which began immediately after receipt of his Ph.D. degree in 1954 from the University of Illinois. He began his research on bimetallic catalysts in the early 1960s. Much of his work in this area has been concerned with developing the concept of highly dispersed "bimetallic cluster" catalysts, and with the application of this concept in the catalytic reforming of petroleum fractions. For his accomplishments in catalysis, Dr. Sinfelt has received a number of awards, including the President's National Medal of Science in 1979. He is a member of the National Academy of Sciences, the National Academy of Engineering, and the American Academy of Arts and Sciences.

CONTENTS

Chapter 1

INTRODUCTION

Scientists have been interested in bimetallic systems as catalysts for many years. For a decade or longer beginning shortly after World War II, much attention was devoted to the use of metal alloys as catalysts to probe the relationship between the catalytic activity of a metal and its electronic structure *(1–4).* One type of alloy which was investigated extensively consisted of a Group VIII and a Group IB metal, for instance, nickel–copper or palladium–gold.

In terms of the energy band theory of electrons in metals *(5),* transition metals such as those of Goup VIII possess *d*-bands whose states are not completely occupied by electrons. By contrast, the *d*-bands of nontransition metals such as those of Group IB are completely filled.

In the case of the nickel–copper system, which can form solid solutions over the whole range of compositions, the substitution of copper atoms for nickel atoms in the metal lattice adds electrons to the system. According to Mott and Jones *(5),* nickel–copper alloys would possess *d*-bands with fewer unoccupied states than would pure nickel. In this view a nickel–copper alloy would possess a single *d*-band rather than separate bands for the two components, and the additional electrons introduced with the copper would lead to an alloy with a more completely filled *d*-band. In this model of the electronic structure of a nickel–copper alloy, no distinction was made between the chemical properties of nickel and copper atoms in the alloy.

To test the hypothesis that the catalytic activity of a Group VIII metal is associated with an unfilled *d*-band, various workers determined reaction rates on alloys such as nickel–copper and palladium–gold as a function of composition. It was reasoned that the extent of filling of the *d*-band would be determined by the composition of the alloy; hence it should be possible to relate catalytic activity to *d*-band vacancies.

Although many studies of this type were conducted, the approach was not very fruitful in elucidating the so-called *electronic factor* in catalysis by metals. The work was based on the premise that the catalytic activity of a metal is determined by the electronic structure of the crystal as a whole. Today, this premise appears questionable. It has been largely supplanted by the view that catalysis is determined by localized properties of surface sites.

Experimental data on chemisorption and catalysis indicate that the different types of atoms in the surface of an alloy, such as nickel and copper, largely retain their chemical identities, although their bonding properties may be modified *(6,7)*. At present the electronic factor in catalysis by metals generally is viewed in terms of localized chemical bonding effects similar to the "ligand" effects of organometallic chemistry *(6)*.

The early work on metal alloys to elucidate the electronic factor devoted much attention to the effect of alloy composition on the rates of a few selected reactions, such as the hydrogenation of ethylene to ethane *(8,9)* and of benzene to cyclohexane *(10)*. Little or no attention was given to fragmentation reactions of hydrocarbons leading to formation of products of lower carbon number than the reactant, such as the hydrogenolysis of ethane to methane. The possibility that alloying effects on fragmentation reactions could be very different from those on hydrogenation (or dehydrogenation) reactions was not appreciated, nor was the selectivity aspect of catalysis by metal alloys in general.

The alloy catalysts used in these early studies were low surface area materials, commonly metal powders or films. The surface areas, for example, were two orders of magnitude lower than that of platinum in a commercial reforming catalyst. Hence these alloys were not of interest as practical catalysts. The systems emphasized in these studies were combinations of metallic elements that formed continuous series of solid solutions, such as nickel–copper and palladium–gold. The use of such systems presumably made it possible to vary the electronic structure of a metal crystal in a known and convenient manner, and thereby to determine its influence on catalytic activity. Bimetallic combinations of elements exhibiting limited miscibility in the bulk were not of interest. Aspects of bimetallic catalysts other than questions related to the influence of bulk electronic structure received little attention in these studies.

In 1963, the author of this monograph began an investigation of bimetallic catalysts in a broader context. The early research was conducted in the Process Research Division and in the Central Basic Research Laboratories of Exxon Research and Engineering Company. Since 1968 it has been conducted in the Exxon Corporate Research Laboratories.

Two items in the research received particular attention: (a) the investigation of selectivity effects in catalysis by such bimetallic materials, and (b) the preparation and characterization of highly dispersed bimetallic catalysts. Two types of bimetallic catalysts were emphasized. One consisted of a combination of atoms of a Group VIII metal and a Group IB metal, and the other consisted of atoms of two different Group VIII metals.

In the work of the author and his associates on bimetallic catalysts comprising various combinations of Group VIII and Group IB metals, it was discovered that the activity of the Group VIII metal for hydrogenolysis reactions of hydrocarbons was decreased markedly by the presence of the Group IB metal *(11–15).* It was shown that the inhibition of hydrogenolysis leads to improved selectivity for alkane isomerization reactions *(11)* and for reactions in which saturated hydrocarbons are converted to aromatic hydrocarbons *(12,14,15).* Interest in bimetallic catalysts increased markedly with the discovery of this selectivity phenomenon.

In the course of this research it was discovered that bimetallic systems of interest were not limited to combinations of metallic elements that are highly miscible in the bulk. For example, the ruthenium–copper system *(14,15),* in which the two components are virtually completely immiscible in the bulk, exhibits selective inhibition of hydrogenolysis similar to that observed with the nickel–copper system. The latter, of course, is well known as a system in which the two components form a complete series of solid solutions.

Since the ability to form bulk alloys was not a necessary condition for a system to be of interest as a catalyst, it was decided not to use the term alloy in referring to bimetallic catalysts in general. Instead, terms such as bimetallic aggregates or *bimetallic clusters* have been adopted in preference to alloys. In particular, *bimetallic clusters* refer to bimetallic entities which are highly dispersed on the surface of a carrier. For systems such as ruthenium–copper, it appears that the two components can interact strongly at an interface, despite the fact that they do not form solid solutions in the bulk. In this system the copper is present at the surface of the ruthenium, much like a chemisorbed species.

Results of studies with catalyst combinations consisting of two metallic elements from Group VIII of the periodic table were also highly interesting. A major new catalyst was discovered for the reforming of petroleum naphtha fractions to produce high antiknock quality gasoline components *(16).* The catalyst, consisting of bimetallic clusters of platinum and iridium dispersed on alumina *(17,18),* was severalfold more active than the prior platinum-on-alumina reforming catalysts and exhibited much better activity maintenance *(16,17).* This discovery was important for production of unleaded gasoline.

The platinum–iridium catalyst developed in our laboratories and a platinum–rhenium catalyst *(19)* developed by the Chevron Company at approximately the same time have largely replaced the older platinum–alumina catalysts in reforming units. The platinum–rhenium system is an example of a combination of elements from Groups VIII and VIIA of the periodic table.

The preparation of highly dispersed bimetallic clusters on conventional carriers, such as silica or alumina, generally has not been a problem. But the characterization of these clusters has presented scientific challenges. In the early studies on ruthenium–copper and osmium–copper clusters *(14),* the use of chemical probes (e.g., chemisorption of hydrogen and carbon monoxide in conjunction with kinetic studies of a test reaction such as the hydrogenolysis of ethane to methane) was invaluable. These probes provided clear evidence of extensive interaction between the copper and either the ruthenium or osmium, thus pointing to the existence of bimetallic clusters. The possibility was thereby discounted that atoms of one of the metallic elements were totally isolated from atoms of the other in clusters which were purely monometallic. In studies of a system such as platinum–iridium, in which the two components exhibit much smaller differences in their chemisorption and catalytic properties than is the case for the ruthenium–copper and osmium–copper systems, the chemical probes are less sensitive but still useful. Physical probes such as X-ray diffraction, Mössbauer effect spectroscopy, and extended X-ray absorption fine structure (EXAFS) are also useful in obtaining information on the structure of bimetallic clusters *(18,20–23).* For very highly dispersed bimetallic clusters, in which virtually all of the metal atoms are surface atoms, X-ray diffraction studies provide little information. In such a case, however, Mössbauer effect studies and EXAFS investigations may still be very useful.

This monograph describes research on bimetallic catalysts conducted at the Exxon Research and Engineering laboratories since the early 1960s. Much of the monograph is concerned with research directed toward the validation and elucidation of the bimetallic cluster concept. Some discussion is devoted also to the technological aspects of these systems, with emphasis on their application for the catalytic reforming of petroleum fractions.

REFERENCES

1. Schwab, G.M. "Alloy catalysts in dehydrogenation." *Disc. Faraday Soc.* **8**: 166–171; 1950.
2. Dowden, D.A. and Reynolds, P. "Some reactions over alloy catalysts." *Disc. Faraday Soc.* **8**: 184–190; 1950.

3. Couper, A. and Eley, D.D. "The parahydrogen conversion on palladium–gold alloys." *Disc. Faraday Soc.* **8**: 172–184; 1950.
4. Best, R.J. and Russell, W.W. "Nickel, copper and some of their alloys as catalysts for ethylene hydrogenation." *J. Amer. Chem. Soc.* **76**: 838–842; 1954.
5. Mott, N.F. and Jones, H. *Theories of the Properties of Metals and Alloys.* London: Oxford University Press; 1936.
6. Sachtler, W.M.H. "Surface composition of alloys in equilibrium." *LeVide.* **164:** 67–70; 1973.
7. Sachtler, W.M.H. and van der Plank, P. "The role of individual surface atoms in chemisorption and catalysis by nickel–copper alloys." *Surface Science.* **18**: 62–79; 1969.
8. Hall, W.K. and Emmett, P.H. "Studies of the hydrogenation of ethylene over copper–nickel alloys." *J. Phys. Chem.* **63**: 1102–1110; 1959.
9. Gharpurey, M.K. and Emmett, P.H. "Study of the hydrogenation of ethylene over homogenized copper–nickel alloy films." *J. Phys. Chem.* **65**: 1182–1184; 1961.
10. Hall, W.K. and Emmett, P.H. "The hydrogenation of benzene over copper–nickel alloys." *J. Phys. Chem.* **62**: 816–821; 1958.
11. Sinfelt, J.H.; Barnett, A.E.; and Dembinski, G.W., inventors; Exxon Research and Engineering Company, assignee. "Isomerization process utilizing a gold–palladium alloy in the catalyst." U.S. Patent 3,442,973. 2 pages. May 6, 1969.
12. Sinfelt, J.H.; Barnett, A.E.; and Carter, J.L., inventors; Exxon Research and Engineering Company, assignee. "Inhibition of hydrogenolysis." U.S. Patent 3,617,518. 4 pages. November 2, 1971.
13. Sinfelt, J.H.; Carter, J.L.; and Yates, D.J.C. "Catalytic hydrogenolysis and dehydrogenation over copper–nickel alloys." *J. Catal.* **24**: 283–296; 1972.
14. Sinfelt, J.H. "Supported 'bimetallic cluster' catalysts." *J. Catal.* **29**: 308–315; 1973.
15. Sinfelt, J.H.; Lam, Y.L.; Cusumano, J.A.; and Barnett, A.E. "Nature of ruthenium–copper catalysts." *J. Catal.* **42**: 227–237; 1976.
16. Sinfelt, J.H. "Esso catalyst based on multimetallic clusters." *Chem. Eng. News.* **50**: 18–19; July 3, 1972.
17. Sinfelt, J.H., inventor; Exxon Research and Engineering Company, assignee. "Polymetallic cluster compositions useful as hydrocarbon conversion catalysts." U.S. Patent 3,953,368. 16 pages. 1976.
18. Sinfelt, J.H. and Via, G.H. "Dispersion and structure of platinum–iridium catalysts." *J. Catal.* **56**: 1–11; 1979.
19. Jacobson, R.L.; Kluksdahl, H.E.; McCoy, C.S.; and Davis, R.W. "Platinum–rhenium catalysts: a major new catalytic reforming development." *Proceedings of the American Petroleum Institute, Division of Refining.* **49**: 504–521; 1969.

20. Garten, R.L. and Sinfelt, J.H. "Structure of Pt–Ir catalysts: Mössbauer spectroscopy studies employing ^{57}Fe as a probe." *J. Catal.* **62**: 127–139; 1980.
21. Sinfelt, J.H.; Via, G.H.; and Lytle, F.W. "Structure of bimetallic clusters. Extended X-ray absorption fine structure (EXAFS) studies of Ru–Cu clusters." *J. Chem. Phys.* **72**: 4832–4844; 1980.
22. Sinfelt, J.H.; Via, G.H.; Lytle, F.W.; and Greegor, R.B. "Structure of bimetallic clusters. Extended X-ray absorption fine structure (EXAFS) studies of Os–Cu clusters." *J. Chem. Phys.* **75**: 5527–5537; 1981.
23. Sinfelt, J.H.; Via, G.H.; and Lytle, F.W. "Structure of bimetallic clusters. Extended X-ray absorption fine structure (EXAFS) studies of Pt–Ir clusters." *J. Chem. Phys.* **76**: 2779–2789; 1982.

ABSTRACT

The potential of bimetallic catalysts to alter reaction selectivity (the rate of one type of chemical reaction relative to another) became evident in the early stages of the research described in this monograph. For catalysts consisting of a Group VIII metal in combination with a Group IB metal, it was discovered that the Group IB metal markedly decreases the activity of the Group VIII metal for hydrogenolysis of hydrocarbons. The hydrogenolysis activity of a Group IB metal is negligible when compared to that of a Group VIII metal. Inhibition of hydrogenolysis results in improved selectivity for those reactions in which saturated hydrocarbons are converted to aromatic hydrocarbons as well as for alkane isomerization reactions. This selectivity effect can be largely rationalized by the hypothesis that hydrogenolysis requires a surface site consisting of an array of adjacent active metal atoms which is larger than that required for hydrogenation, dehydrogenation, or isomerization.

Chapter 2

SELECTIVITY ASPECTS OF BIMETALLIC CATALYSTS

During our research on bimetallic catalysts, it was evident very early that the activities of a metal catalyst for different reactions could be altered to markedly different degrees by the incorporation of a second metallic element into the catalyst. The discussion begins with a brief review of our early exploratory studies of this selectivity phenomenon, with emphasis on hydrocarbon reactions such as isomerization of alkanes, aromatization of alkylcyclopentanes, dehydrogenation of cyclohexanes to aromatic hydrocarbons, and hydrogenolysis of alkanes and cycloalkanes.

This review is followed by a consideration of some of the features characteristic of hydrocarbon reactions on catalysts comprising individual metals from Groups VIII and IB of the periodic table. Finally, the activities of a series of unsupported nickel–copper alloys for hydrogenolysis and dehydrogenation reactions are discussed. These latter studies were made to obtain information on the selectivity phenomenon with bimetallic catalysts of known structure. The nickel–copper alloys were characterized by a variety of chemical and physical probes.

The work with nickel–copper alloys led to a better understanding of the selectivity phenomenon than did the original exploratory studies on supported bimetallic catalysts, since the supported catalysts were difficult to characterize with techniques available at that time. Nevertheless, the early exploratory studies were important in disclosing the selectivity phenomenon and in providing incentive to conduct further research.

2.1 RESULTS OF EARLY EXPLORATORY STUDIES

In the conversion of *n*-heptane over a series of palladium–gold catalysts in which alumina was employed as a carrier, the presence of gold decreased the extent of formation of C_1–C_6 alkanes and simultaneously increased the formation of C_7 isomers, consisting of methylhexanes and dimethylpentanes. The data in Table 2.1 were taken from our 1969 patent *(1),* the application for which was filed in 1966. The catalysts were prepared by contacting alumina with aqueous solutions of palladium and gold compounds. The resulting materials were dried and reduced in hydrogen at 454°C. Although the presence of the gold improves the isomerization selectivity of the palladium catalyst, gold by itself on alumina is not an effective catalyst for the reaction.

The presence of gold with the palladium also improves the selectivity of such catalysts for the conversion of methylcyclopentane to benzene. In this case, reactions leading to the formation of C_1–C_6 alkanes as a result of the rupture of the five-membered ring structure in methylcyclopentane are inhibited relative to the aromatization reaction. Data from our 1973 patent *(2),* the application for which was filed in 1971, are shown in Table 2.2.

Palladium–gold on alumina catalysts are examples of bifunctional catalysts *(3,4),* in which the alumina carrier itself possesses acidic sites that are involved in the reactions. This property of the system impedes a detailed understanding of the effect of the gold in suppressing the formation of C_1–C_6 alkanes. However, when alkanes or cycloalkanes are contacted with simpler metal catalysts, specifically, metals which are either unsupported or supported on a relatively nonacidic carrier such as silica, a similar marked suppression in the formation of products of lower carbon number than the reactants is observed when a Group IB metallic element is incorporated with a Group VIII metal. Examples are given in our 1971 patent *(5),* for which the

Table 2.1 Conversion[a] of *n*-Heptane over Pd–Au/Al_2O_3 Catalysts *(1)*

		% Selectivity to			
Catalyst[b]	% Conversion	C_1+C_2	C_3–C_6	Toluene	C_7 Isomers
0.6% Pd	45.5	2.9	37.9	7.6	50.0
0.6% Pd, 0.1% Au	46.5	2.2	20.0	9.0	66.9
0.6% Pd, 0.6% Au	34.7	1.6	10.9	5.7	78.7
0.6% Pd, 1.2% Au	50.3	1.6	11.0	4.9	81.3
0.6% Au	10.5	15.0	60.5	<0.1	24.5

[a]Conditions: 454°C, 14.6 atm, $H_2/nC_7 = 5/1$; weight hourly space velocity of *n*-heptane = 20; product analyses after 15 min on stream.

[b]Alumina carrier. Percentages for metals are by weight based on total catalyst mass.

Table 2.2 Selectivity of Methylcyclopentane Conversion[a] on Pd and Pd–Au Catalysts *(2)*

Catalyst[b]	% Conversion	% Selectivity to Benzene
0.6% Pd	21.6	57.2
0.6% Pd, 0.3% Au	25.7	69.4

[a]Conditions: 454°C, 14.6 atm; 5/1 mole ratio of hydrogen to methylcyclopentane.

[b]Alumina carrier. Percentages for metals are by weight on total catalyst mass.

application was filed in 1969. Data from this patent showing the effect of copper and silver on the selectivity of ruthenium for the conversion of cyclohexane to benzene are presented in Table 2.3. The metals were dispersed on silica in these catalysts. On pure ruthenium, cyclohexane undergoes extensive hydrogenolysis to low molecular weight alkanes, primarily methane, in addition to dehydrogenation to benzene. The inclusion of either copper or silver with ruthenium leads to marked inhibition of the hydrogenolysis reaction, with accompanying substantial improvement in selectivity for benzene formation. Catalysts containing only copper or silver on silica exhibit very low activity. Selectivity effects similar to those shown in Table 2.3 are observed with silica-supported nickel–copper and cobalt–copper catalysts, as demonstrated by additional data given in the patent.

Table 2.3 Effect of Copper and Silver on the Selectivity[a] of Ruthenium for Cyclohexane Conversion *(5)*

Catalyst[b]	% Conversion	% Selectivity to Benzene
1% Ru	15.1	76.0
1% Ru, 0.1% Cu	9.7	83.5
1% Ru, 1% Cu	10.4	94.0
5% Cu	<0.5	—
1% Ru, 1% Ag	10.7	87.9
1% Ag	<0.1	—

[a]Conditions: 343°C, 1 atm; mole ratio of hydrogen to cyclohexane = 5/1; weight hourly space velocity of cyclohexane = 25 grams per hour per gram of catalyst.

[b]Silica carrier. Percentages for metals are by weight based on total catalyst mass.

In general, inclusion of a Group IB metal with a Group VIII metal markedly decreases the hydrogenolysis activity of the latter but has a much smaller effect on the activity for reactions such as the dehydrogenation and isomerization of hydrocarbons *(1,2,5–8)*. These observations have been supported by the work of other investigators *(9–11)*.

It is particularly interesting that selectivity effects have been observed even for bimetallic systems that do not form solid solutions in the bulk. For example, ruthenium and copper are virtually completely immiscible in the bulk *(12)*, yet the catalytic data provide evidence of marked interaction between these two components in ruthenium–copper catalysts. This phenomenon is discussed in greater detail in Chapters 3 and 4.

The results of our early exploratory studies on bimetallic catalysts clearly demonstrated that selectivity effects are important in hydrocarbon reactions on bimetallic catalysts. To obtain insight into the factors involved in the selectivity effects, it is useful at this point to inquire into the nature of the hydrocarbon reactions under consideration.

2.2 NATURE OF SELECTED HYDROCARBON REACTIONS ON METALS

The consideration of rates and mechanistic aspects of reactions such as hydrogenolysis, dehydrogenation, and isomerization provides a basis for interpreting selectivity data on bimetallic catalysts.

2.2.1 Hydrogenolysis

Hydrogenolysis of hydrocarbons involves the rupture of carbon–carbon bonds and the formation of carbon–hydrogen bonds. The simplest hydrogenolysis reaction of a hydrocarbon is the conversion of ethane to methane

$$C_2H_6 + H_2 \rightarrow 2CH_4$$

This reaction has been studied in detail over a number of metals *(13–20)*. The reaction may be dissected into a sequence of steps, where the symbol (ads) signifies an adsorbed species *(21–23)*

$$C_2H_6 \rightleftarrows C_2H_x\text{(ads)} + a\,H_2$$

$$C_2H_x\text{ (ads)} \rightarrow \text{adsorbed } C_1 \text{ fragments}$$

The quantity a is equal to $(6-x)/2$. Ethane is first chemisorbed with dissociation of carbon–hydrogen bonds, ultimately yielding a hydrogen-deficient surface species, C_2H_x. The latter undergoes carbon–carbon scission to yield adsorbed C_1 fragments (e.g., adsorbed CH or CH_2) which are subsequently hydrogenated to methane.

Figure 2.1 compares the catalytic activities of all of the metals of Group VIII and of rhenium in Group VIIA for ethane hydrogenolysis. Three separate fields represent the metals of the first, second, and third transition series *(22,23).* The Group IB metals (copper, silver, gold), for which data are not shown, are much less active than the least active of the Group VIII metals *(22,23).*

The most complete data available are for metals of the third transition series, for which the hydrogenolysis activity reaches a maximum value with osmium. From osmium to platinum, the activity decreases by eight orders of magnitude. A similar variation is observed from ruthenium to palladium in the second transition series. In the first transition series, the pattern of variation of hydrogenolysis activity is different. Maximum activity is found in the third, rather than the first, subgroup within Group VIII. This difference is somewhat analogous to known chemical differences between elements of the first transition series on the one hand and the corresponding elements of the second and third transition series on the other *(24).*

The pattern of variation of catalytic activity of the metals of the second and third transition series for the hydrogenolysis of ethane is also observed in *n*-heptane hydrogenolysis *(22,25,26),* as shown in Figure 2.2. Products of the hydrogenolysis of *n*-heptane are C_1–C_6 alkanes, the relative amounts depending on the metal. Absolute rates of hydrogenolysis are approximately two to three orders of magnitude higher for *n*-heptane than for ethane *(25).* Reasons for this difference are discussed elsewhere *(26).*

Included with the plots of hydrogenolysis activities of the metals in Figure 2.1 are plots of percentage *d*-character of the metallic bond, a quantity introduced by Pauling *(27)* as a measure of the extent of participation of *d*-orbitals in the bonding between atoms in a metal lattice. For the metals within a given transition series, the patterns of variation of hydrogenolysis activity and percentage *d*-character from one metal to another are similar. However, the hydrogenolysis activities of iron, cobalt, and nickel in the first transition series are comparable to those of metals with significantly higher *d*-character in the second and third transition series. Thus percentage *d*-character alone is not adequate for characterizing the catalytic activity of transition metals for hydrogenolysis *(22).*

The specific activity of a metal catalyst for the hydrogenolysis of an alkane

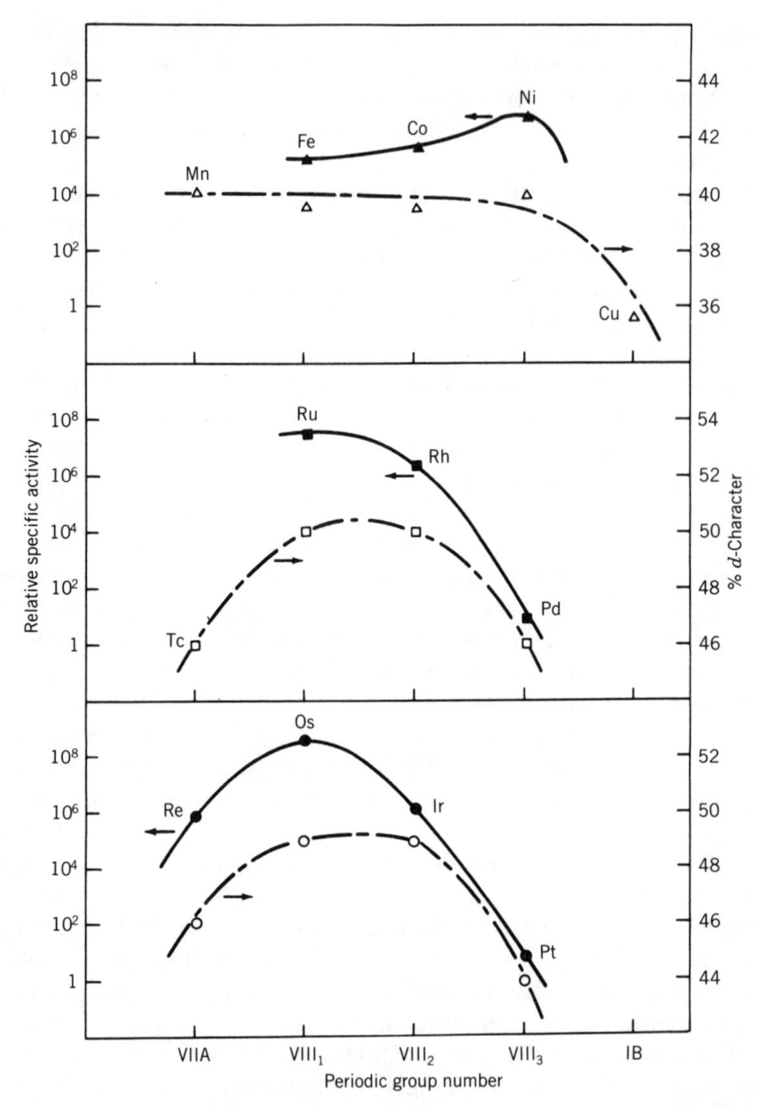

Figure 2.1 Catalytic activities of metals for ethane hydrogenolysis in relation to percentage *d*-character of the metallic bond. Closed points are activities at 205°C, and ethane and hydrogen pressures of 0.030 and 0.20 atm, respectively; open points are percentage *d*-character *(22,23)*. (Reprinted with permission from Academic Press, Inc.)

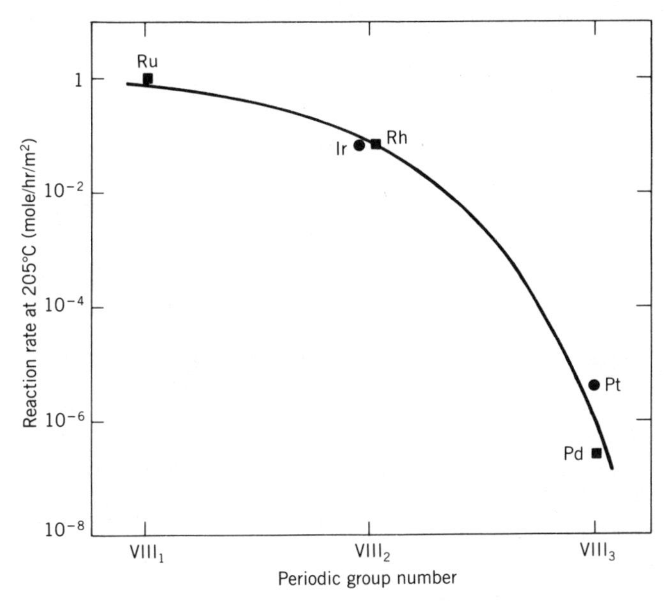

Figure 2.2 Catalytic activities of Group VIII noble metals for *n*-heptane hydrogenolysis *(22,25,26)*. (Conditions: 205°C, 1 atm, H_2/nC_7 mole ratio = 5/1.) (Reprinted with permission from Academic Press, Inc.)

or cycloalkane depends in general on the state of dispersion of the metal; that is, the catalytic activity per surface metal atom for a large metal crystal differs from that for a very small metal crystallite or cluster. Data illustrating this point are shown in Figure 2.3 for the hydrogenolysis of cyclohexane over a series of ruthenium catalysts varying in dispersion by two orders of magnitude *(28)*.

Dispersion is defined as the ratio of surface atoms to total atoms in the metal crystallites, and it is determined from chemisorption measurements *(26,29)*. A typical hydrogen chemisorption isotherm is shown in Figure 2.4 for a silica-supported ruthenium catalyst containing 5 wt% ruthenium. The quantity H/Ru in the right-hand ordinate of the figure is the ratio of the number of hydrogen atoms adsorbed to the number of ruthenium atoms in the catalyst. The catalyst was treated with a stream of hydrogen in an adsorption cell at 500°C, after which the cell was evacuated and cooled to room temperature for the determination of the isotherm. The adsorption is

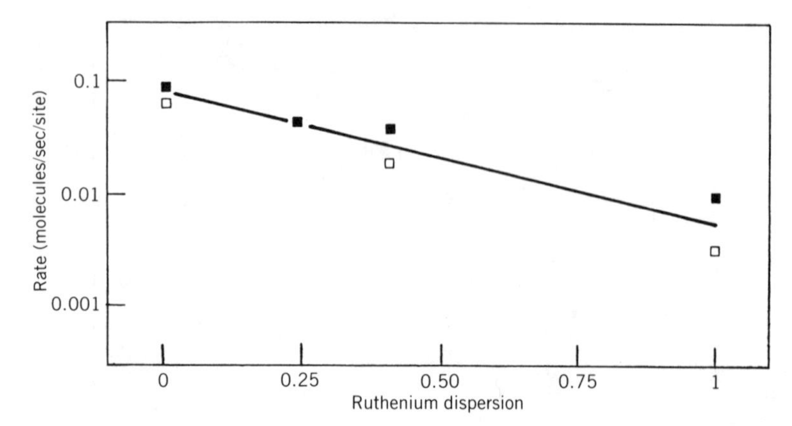

Figure 2.3 The effect of the degree of ruthenium dispersion on its catalytic activity for hydrogenolysis of cyclohexane to lower carbon number alkanes *(28)*. (Solid points represent rates obtained after 10–20 minutes of exposure of the catalyst to reactants; open points represent data obtained after 2 hours of exposure.) (Reprinted with permission from Academic Press, Inc.)

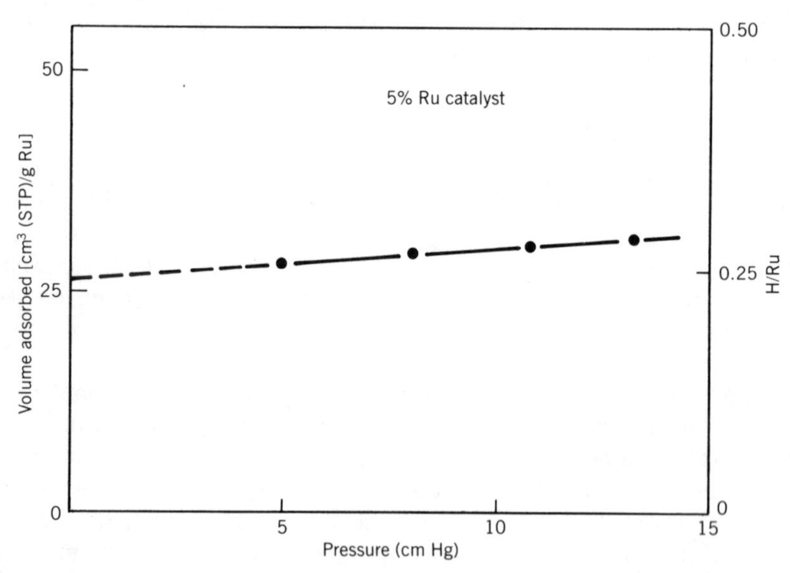

Figure 2.4 Typical hydrogen chemisorption isotherm at room temperature for a silica-supported ruthenium catalyst containing 5 wt% ruthenium *(28)*. (Reprinted with permission from Academic Press, Inc.)

only slightly pressure dependent over the range of pressures employed in obtaining the data. The small pressure dependence is associated with a weakly bound fraction of the total chemisorption. An estimate of the strongly bound fraction is commonly made by extrapolating the data back to zero pressure, since saturation with regard to the strongly chemisorbed component is attained at equilibrium pressures very much lower than those corresponding to the measured isotherm. An alternative method of estimating the amount of strongly chemisorbed hydrogen, which is considered later, involves determination of the amount of hydrogen retained on the catalyst after evacuation of the adsorption cell at room temperature *(6).* The strongly chemisorbed hydrogen determined in this manner is generally somewhat lower than the amount determined by extrapolation of the original isotherm to zero pressure.

In the determination of the dispersion values in the abscissa of Figure 2.3, the method of extrapolation to zero pressure was employed, and an adsorption stoichiometry of one hydrogen atom per surface ruthenium atom was adopted. The specific activity of ruthenium for cyclohexane hydrogenolysis decreases with increasing dispersion, declining by one order of magnitude as the dispersion increases by two orders of magnitude. The products of the hydrogenolysis reaction are lower carbon number alkanes, predominantly methane.

In the interpretation of data on the hydrogenolysis activities of metals, it has commonly been hypothesized that the chemisorbed hydrocarbon intermediate forms a number of bonds with the metal surface, and that a surface site consisting of a single active metal atom is not adequate *(6,9).* Finding a suitable array of surface atoms to accommodate such a chemisorbed intermediate presents no difficulty on a large metal crystal where most of the surface atoms are present in the faces of the crystal. On very small crystallites, however, a large fraction of the surface atoms exists at edges and corners. If the required array comprises a large number of metal atoms, the probability of finding such an array in the surface may be substantially lower than on large crystals. If such an array of active metal atoms is required for hydrogenolysis, it might be expected that the specific hydrogenolysis activity of highly dispersed metal clusters would be lower than that of large crystals, as observed for ruthenium. It might also be expected that atoms of an inactive second element dispersed on top of the surface, or within the surface layer, would greatly decrease the availability of the required arrays and hence decrease hydrogenolysis activity strikingly. As will be seen, this expectation is indeed borne out in hydrogenolysis studies on bimetallic catalysts.

2.2.2 Hydrogenation and Dehydrogenation Reactions

In comparison with alkane hydrogenolysis reactions, hydrogenation and dehydrogenation reactions of hydrocarbons on metals exhibit a much smaller range of variation in rate (per unit surface area or per surface atom) from one metal to another. They also exhibit smaller effects of the degree of metal dispersion on the rate per surface atom. Thus, while rates of ethane hydrogenolysis vary by about eight orders of magnitude among the Group VIII metals *(22,23),* rates of hydrogenation of benzene to cyclohexane and of ethylene to ethane vary by about two and three orders of magnitude, respectively *(30).* For cyclohexane dehydrogenation to benzene, the available data indicate a similar lower range of variation of rates among the metals *(5–8,31).* Moreover, rates of benzene hydrogenation and cyclohexane dehydrogenation per surface metal atom are essentially unchanged when metal dispersion is varied by an order of magnitude *(31–34).* As a consequence of the small effect of metal dispersion, these reactions have been characterized as structure-insensitive, the terminology having been introduced by Boudart *(29).*

In the case of reactions which have been identified as structure-insensitive, the view is commonly held that the surface site need not consist of a large array of active metal atoms. Perhaps a site containing a single metal atom is adequate *(35).* As a consequence, the incorporation of atoms of an inactive second component within or on the surface of the active metal should have a substantially smaller effect in decreasing the number of catalytically active sites for this class of reactions than for reactions in which the required sites consist of large arrays of active metal atoms. For example, a 50% random replacement of, or coverage of, the metal atoms in a surface layer by inactive atoms would decrease the probability of finding a site consisting of a single atom of active metal by a factor of two, whereas the probability of finding a site consisting of an array of four adjacent atoms of active metal would decrease by a factor of sixteen. Thus hydrogenation or dehydrogenation reactions would not be expected to behave like hydrogenolysis on bimetallic catalysts consisting of an active Group VIII metal in combination with an inactive Group IB metal. This line of reasoning leads to the expectation that the presence of the Group IB metal will markedly inhibit the hydrogenolysis activity of the Group VIII metal, thus rendering it more selective for hydrogenation or dehydrogenation reactions.

While considerations of surface composition and structure alone provide a basis for anticipating selectivity effects, the possibility cannot be dismissed that such effects are also influenced by electronic interactions between the two metal components of the catalyst. This matter is discussed in more detail in Section 2.3.

2.2.3 Isomerization Reactions

Isomerization reactions of alkanes and cycloalkanes occur very readily on bifunctional catalysts containing both metal and acidic sites *(3,4),* the latter being associated with the carrier employed for the metal. This mode of reaction is very important for the catalysts used in commercial reforming, which will be discussed in Chapter 5. In such bifunctional catalysts, the metal and acidic sites catalyze different steps in the reaction sequence.

It is also well established that isomerization reactions of alkanes and cycloalkanes can occur on metal catalysts which do not contain separate acidic sites *(25,36,37).* Such catalysts have been termed monofunctional catalyst systems. In studies of the conversion of *n*-hexane on platinum metal films of variable thickness, including ultrathin films consisting of microcrystallites with sizes of 20 Å or lower, it has been reported that the conversion to isomerization products (2-methylpentane and 3-methylpentane) relative to hydrogenolysis products (C_1–C_5 alkanes) is very much higher on the ultrathin films than on thick films *(37).* From these results, it was concluded that the isomerization reaction could occur on sites consisting of single metal atoms, such as the corner atoms of a crystal. The fraction of metal atoms present at corners is much higher in the microcrystallites constituting the ultrathin films than in the thick films not characterized by this type of microstructure. The improved selectivity to isomerization in the case of the ultrathin films then results because hydrogenolysis requires a site consisting of a large array of active metal atoms of the type present in the face of a crystal. As a consequence of this conclusion, it might be expected, as in hydrogenation and dehydrogenation reactions, that selectivity to isomerization on an active Group VIII metal surface would be enhanced by the random incorporation of atoms of an inactive second component at the surface (either within, or on top of, the surface layer).

In considering the results of Table 2.1, the decreased hydrogenolysis activity of the palladium may be attributed to the influence of the gold in breaking up the types of arrays of active metal atoms required for hydrogenolysis. This effect would account for the improved isomerization selectivity of the palladium–gold catalysts, but the bifunctional nature of the catalyst precludes the possibility of simply attributing the different observed effect of the gold on isomerization to the ability of the latter reaction to occur on single metal atom sites. However, in studies of hexane isomerization on unsupported nickel–

copper alloys, which do not contain a separate acidic function, such an interpretation is reasonable for accounting for the effect of the copper in improving isomerization selectivity *(10)*.

2.3 STUDIES ON NICKEL–COPPER ALLOYS

Nickel–copper alloys provide a good example of a bimetallic catalyst system in which the variation of catalytic activity with composition depends markedly on the type of reaction, thus leading to substantial selectivity effects. The catalysts to be considered here are alloy powders with a surface area of approximately 1 m^2/g *(6)*. Approximately one atom out of a thousand is a surface atom in such catalysts.

Each alloy catalyst was prepared by a coprecipitation method in which ammonium bicarbonate was added to an aqueous solution of nickel and copper nitrates. The resulting precipitate was dried and heated in air at 370°C to form a mixture of nickel and copper oxides. The mixed oxides were then reduced in hydrogen in several stages over a range of temperatures to produce the nickel–copper alloy. The reduction was completed at 400°C.

Before data are considered on rates of reactions occurring on the alloys, information is presented on the properties of the nickel–copper catalysts themselves. The lattice constants of the alloys were determined as a function of the composition by means of X-ray diffraction, as shown in Figure 2.5. Data on a number of metallurgical preparations taken from the work of Coles *(38)* are included for comparison. The variation of lattice constant with composition is approximately linear.

Data on the magnetic properties of the nickel–copper catalysts are also of interest. The variation of the magnetization with composition is shown in Figure 2.6. The points in the figure are data on the same nickel–copper catalysts on which the X-ray diffraction data in the previous figure were obtained. Magnetization data on metallurgical preparations reported by Ahern et al. *(39)* are included for comparison. As expected, the incorporation of copper with the nickel decreases the magnetization of the latter. The data on lattice constants and magnetic properties indicate that the nickel–copper catalyst preparations have bulk properties similar to those of metallurgical alloys of nickel and copper.

Evidence that the surface compositions of the nickel–copper catalysts are different from the bulk compositions is provided by data on the chemisorption of hydrogen on the catalysts at room temperature. This evidence is based on the observation that strong chemisorption of hydrogen does not occur

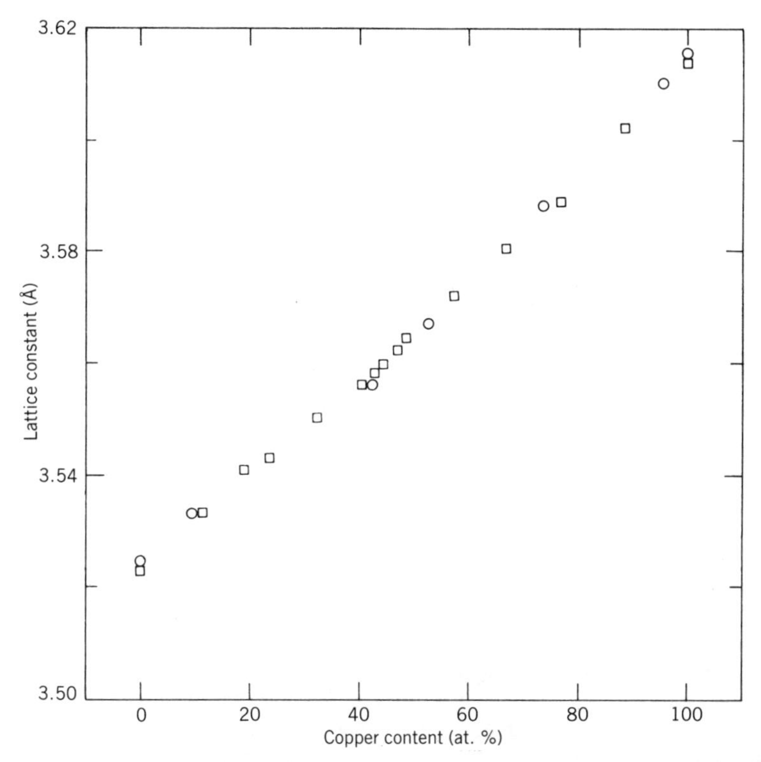

Figure 2.5 Lattice constants for nickel–copper alloys as a function of composition *(6)*. [Circles are data for low surface area (~ 1 m^2/g) catalysts; squares are data of Coles *(38)* on metallurgical specimens.] (Reprinted with permission from Academic Press, Inc.)

on copper. Such a difference in composition can be understood in general terms by a principle first put forth by Gibbs: If accumulation of one of the components in the surface serves to lower the surface energy of the system, the surface will then be enriched in this particular component. It is found that the addition of only a few percent of copper to nickel decreases the amount of strongly chemisorbed hydrogen severalfold, which suggests that the concentration of copper in the surface is much greater than in the bulk. Hydrogen chemisorption data on nickel–copper alloys are given in Figure 2.7, in which the amount of strongly chemisorbed hydrogen is shown as a function of the copper content of the alloy *(6,23)*. Here, strongly chemisorbed hydrogen refers to the amount retained on the catalyst after 10 minutes evacuation of

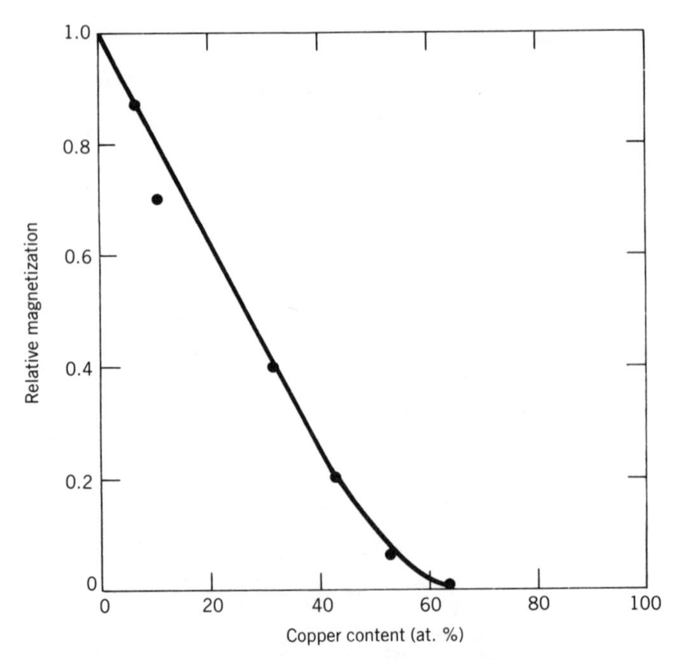

Figure 2.6 Magnetization of nickel–copper alloys as a function of composition *(6)*. [Points are data for the catalysts of Figure 2.5; curve represents data of Ahern et al. *(39)* on metallurgical specimens.] (Reprinted with permission from Academic Press, Inc.)

the adsorption cell to a pressure of approximately 10^{-6} torr at room temperature following completion of an adsorption isotherm.

The data on strongly chemisorbed hydrogen in Figure 2.7 were obtained from adsorption isotherms such as those shown in Figure 2.8 for nickel and copper catalysts and for a nickel–copper alloy catalyst. In each of the three fields of Figure 2.8 isotherms are shown for total hydrogen adsorption and for weakly adsorbed hydrogen. The isotherms are determined as follows: After the catalyst is reduced in situ in the adsorption cell at 450°C with flowing hydrogen, the cell is evacuated and cooled to room temperature for the determination of the isotherm for total adsorption. When this isotherm is completed, the weakly adsorbed hydrogen is removed from the catalyst by evacuating the adsorption cell at room temperature for 10 minutes to a pressure of approximately 10^{-6} torr. A second isotherm is then run at room temperature. This isotherm constitutes the weakly adsorbed hydrogen. The differ-

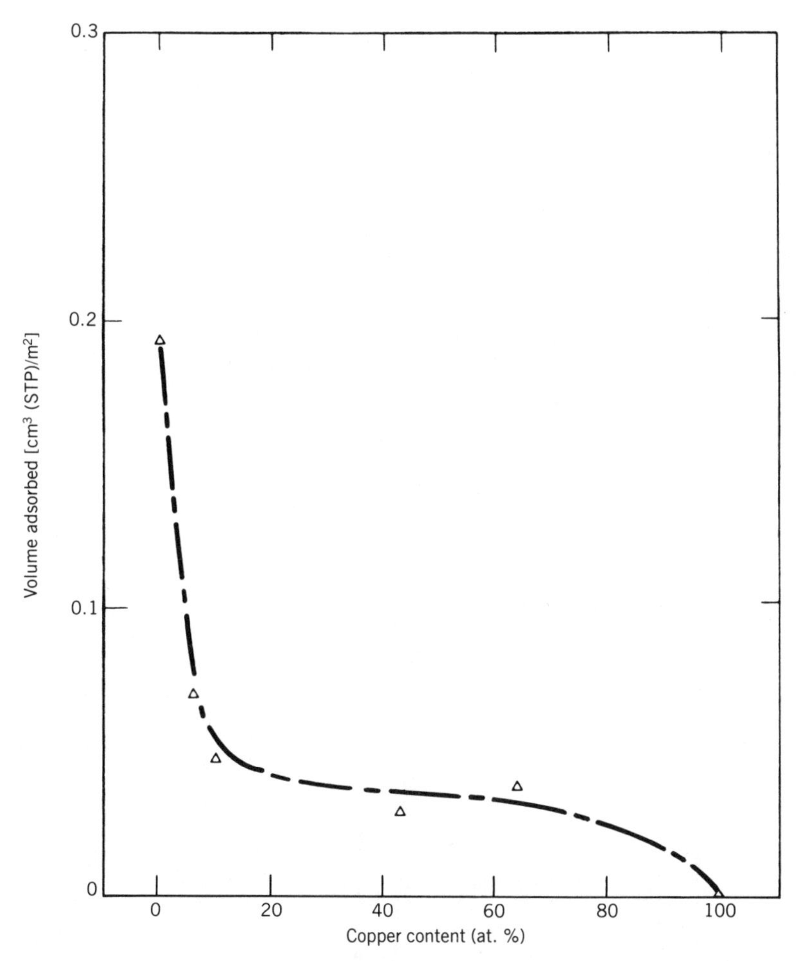

Figure 2.7 Chemisorption of hydrogen on nickel–copper catalysts at room temperature as a function of composition *(6,23)*. (Reprinted with permission from Academic Press, Inc.)

ence between the total adsorption and the weakly adsorbed hydrogen is strongly chemisorbed hydrogen. Note that the total hydrogen adsorption and the weakly adsorbed hydrogen are indistinguishable for pure copper. Thus a strongly chemisorbed fraction is not observed for the copper catalyst.

The data of Figure 2.7 suggest that copper is the predominant component

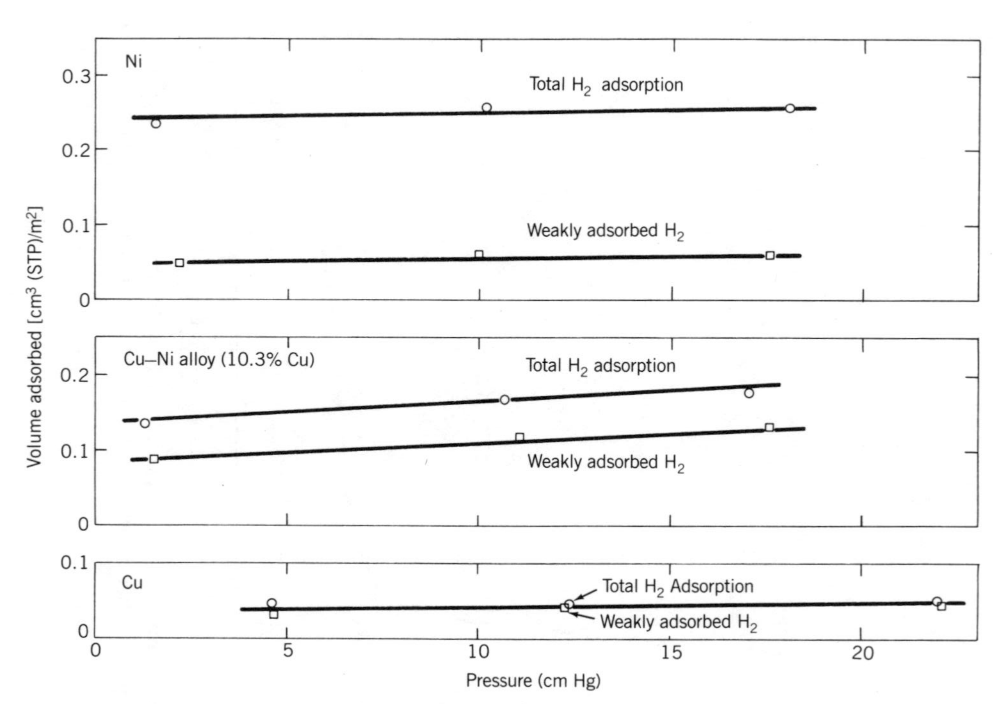

Figure 2.8 Isotherms for total hydrogen adsorption (circles) and weakly adsorbed hydrogen (squares) at room temperature on unsupported nickel and copper catalysts and on a nickel–copper alloy catalyst *(6)*. (Reprinted with permission from Academic Press, Inc.)

in the surface of nickel–copper alloys containing as little as 5 at.% copper overall. Similar data on hydrogen chemisorption on nickel–copper catalysts have been reported by others *(40,41).* The findings are consistent with data on surface compositions obtained by Helms using Auger spectroscopy *(42).* In considering the surface composition of alloys, one must be aware that the nature of the gas in contact with an alloy surface can have a pronounced effect on the composition *(43,44).* Thus if a gas interacts sufficiently strongly and selectively with one of the components of an alloy, the surface tends to be enriched in that particular component even if the other component is the predominant surface species in an inert atmosphere.

In Figure 2.9 the effect of copper on the catalytic activity of nickel is shown for two reactions, the hydrogenolysis of ethane to methane and the dehydrogenation of cyclohexane to benzene, the latter represented by the equation

$$C_6H_{12} \rightarrow C_6H_6 + 3H_2$$

The activities are reaction rates per unit surface area at 316°C. Ethane hydrogenolysis activities were obtained at ethane and hydrogen partial pressures of 0.030 and 0.20 atm, respectively. Cyclohexane dehydrogenation activities were determined at a cyclohexane partial pressure of 0.17 atm and a hydrogen partial pressure of 0.83 atm.

The effect on activity for the dehydrogenation reaction is very different from that for the hydrogenolysis reaction. In the case of ethane hydrogenolysis, adding only 5 at.% copper to nickel decreases catalytic activity by three orders of magnitude. Further addition of copper continues to decrease the activity. However, the activity of nickel for dehydrogenation of cyclohexane is affected very little over a wide range of composition, and actually increases on addition of the first increments of copper to nickel. Only as the catalyst composition approaches pure copper is a marked decline in catalytic activity observed.

To interpret the ethane hydrogenolysis data, we recall the hypothesis of a hydrogen-deficient surface intermediate, C_2H_x, which is bonded to more than one metal atom in the surface. If the composition of the intermediate were to correspond to C_2H_2, for example, we might visualize a structure of the following form:

```
H—C—C—H
  **  **
```

The asterisks represent bonding of carbon atoms to active metal surface atoms. Such an intermediate would require sites comprising *multiplets* of

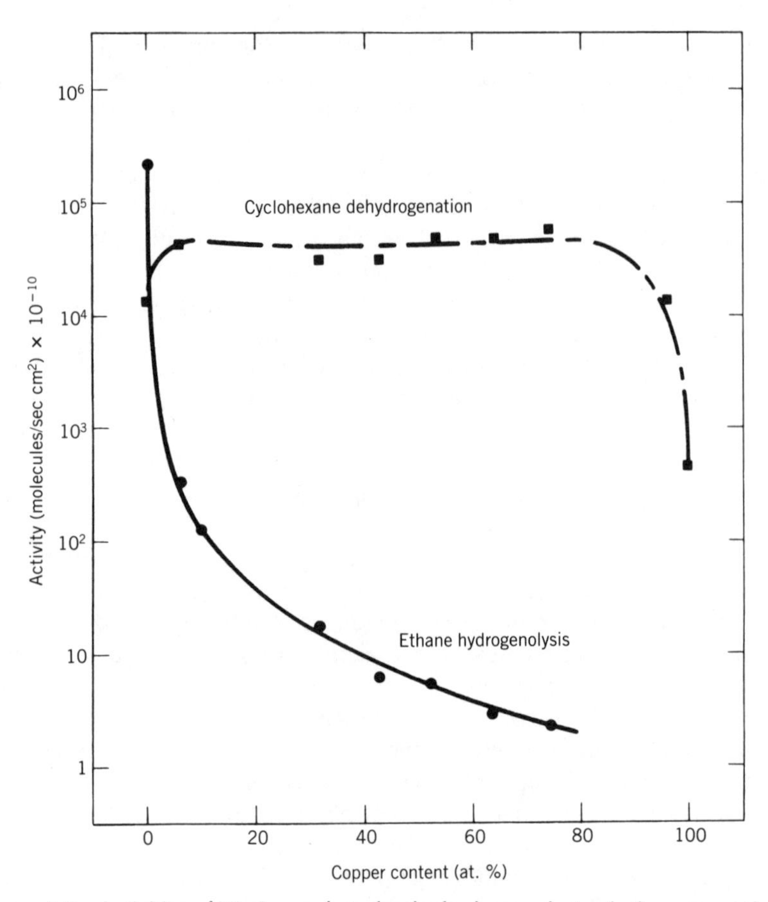

Figure 2.9 Activities of Ni–Cu catalysts for the hydrogenolysis of ethane to methane and the dehydrogenation of cyclohexane to benzene *(6)*. (Reprinted with permission from Academic Press, Inc.)

adjacent active metal atoms. The term multiplet is taken from the work of Balandin *(45)*. More recently, others have used the term *ensemble (9,46)*. If the active metal atoms are diluted with inactive metal atoms in the surface, the concentration of active multiplets will decline sharply. For the nickel–copper system, in which the inactive copper atoms concentrate strongly in the surface, the addition of only a few percent of copper to nickel will result in a markedly lower concentration of multiplet nickel atom sites.

While such reasoning can account for a large inhibiting effect of copper on the hydrogenolysis activity of nickel, it is difficult to dismiss the possibility that electronic interaction between copper and nickel may affect the catalysis. In view of the low ability of copper relative to nickel to chemisorb a variety of hydrocarbons, it reasonably might be expected that addition of copper to nickel in an alloy would decrease the strength of adsorption of hydrocarbon species on the surface. In ethane hydrogenolysis, the strength of bonding between the two carbon atoms in the chemisorbed intermediate might be expected to vary inversely with the strength of bonding of the carbon atoms to the metal. Rupture of the carbon–carbon bond would then be inhibited by a decrease in the strength of adsorption accompanying addition of copper to nickel. If carbon–carbon rupture is rate limiting, the rate of hydrogenolysis should then decrease.

For cyclohexane dehydrogenation, a site consisting of a multiplet of active nickel atoms may not be required. While this would account for the absence of a steep decline in activity as copper is added to nickel, it does not explain why copper-rich alloys have dehydrogenation activities as high or higher than that of pure nickel *(35)*. However, if the activity is controlled by a step whose rate is inversely related to the strength of adsorption (e.g., desorption of the benzene product), the addition of copper to nickel could increase the activity of a catalytic site and compensate for a decrease in the number of such sites. Over the range of composition from pure nickel to pure copper, however, it is likely that the rate-determining step changes. For pure copper, the chemisorption of the cyclohexane itself may be limiting.

At the same time that our work on ethane hydrogenolysis and cyclohexane dehydrogenation on nickel–copper alloys was published, a paper by Ponec and Sachtler on the reactions of cyclopentane with deuterium appeared *(9)*. These workers reported data on the rates of formation of deuterocyclopentanes via exchange, and of CD_4 by hydrogenolysis. The exchange reaction occurred at about the same rate (per surface nickel atom) on nickel–copper alloys as on pure nickel, while the rate of formation of CD_4 was substantially decreased.

The behavior of the exchange reaction thus resembled that of the cyclohexane dehydrogenation reaction in our work. Moreover, the results on formation of CD_4 from cyclopentane resembled our results on the production of CH_4 from ethane. Thus the work of both groups indicated that reactions of carbon–hydrogen bonds on nickel–copper alloys behave very differently from reactions involving rupture of carbon–carbon bonds. As we noted already, this selectivity phenomenon is observed in general for bimetallic catalysts in which a Group IB metal is incorporated with a Group VIII metal.

Other studies from the Netherlands workers on reactions of *n*-hexane *(10)* and of methylcyclopentane *(47)* on nickel–copper alloys yielded results which resembled closely the results we had observed for supported palladium–gold catalysts *(1,2),* summarized in Tables 2.1 and 2.2.

In all of these studies, the presence of the Group IB metal, copper or gold, with the Group VIII metal, nickel or palladium, led to inhibition of hydrogenolysis on the Group VIII metal. As a consequence, selectivities for isomerization of the *n*-alkanes to branched alkanes, and for aromatization of methylcyclopentane to benzene, were improved. The studies of the Netherlands workers on nickel–copper alloys, like those conducted in our laboratories on the same system, eliminated the possibility of complications caused by the presence of a carrier and were therefore useful for clarifying the selectivity phenomenon.

REFERENCES

1. Sinfelt, J.H.; Barnett, A.E.; and Dembinski, G.W., inventors; Exxon Research and Engineering Company, assignee. "Isomerization process utilizing a gold–palladium alloy in the catalyst." U.S. Patent 3,442,973. 2 pages. May 6, 1969.
2. Sinfelt, J.H.; Barnett, A.E.; and Carter, J.L., inventors; Exxon Research and Engineering Company, assignee. "Plural stage reforming with a palladium catalyst in the initial stage." U.S. Patent 3,769,201. 3 pages. October 30, 1973.
3. Sinfelt, J.H. "Bifunctional catalysis." *Advan. Chem. Eng.* **5**: 37–74; 1964.
4. Sinfelt, J.H. "Catalytic reforming of hydrocarbons." In: Anderson, John R. and Boudart, Michel, eds. *Catalysis–Science and Technology.* Vol. 1. Berlin, Heidelberg: Springer-Verlag; 1981; p. 257–300.
5. Sinfelt, J.H.; Barnett, A.E.; and Carter, J.L., inventors; Exxon Research and Engineering Company, assignee. "Inhibition of hydrogenolysis." U.S. Patent 3,617,518. 4 pages. November 2, 1971.
6. Sinfelt, J.H.; Carter, J.L.; and Yates, D.J.C. "Catalytic hydrogenolysis and dehydrogenation over copper–nickel alloys." *J. Catal.* **24**: 283–296; 1972.
7. Sinfelt, J.H. "Supported 'bimetallic cluster' catalysts." *J. Catal.* **29**: 308–315; 1973.
8. Sinfelt, J.H.; Lam, Y.L.; Cusumano, J.A.; and Barnett, A.E. "Nature of ruthenium–copper catalysts." *J. Catal.* **42**: 227–237; 1976.
9. Ponec, V. and Sachtler, W.M.H. "The reactions between cyclopentane and deuterium on nickel and nickel–copper alloys." *J. Catal.* **24**: 250–261; 1972.
10. Ponec, V. and Sachtler, W.M.H. "Reactions of hexane isomers on nickel/copper alloys." In: Hightower, J.W., ed. *Proceedings of the Fifth International Congress on Catalysis.* Vol. 1. Amsterdam: North–Holland; 1973; p. 645–652.

11. Beelen, J.M.; Ponec, V.; and Sachtler, W.M.H. "Reactions of cyclopropane on nickel and nickel–copper alloys." *J. Catal.* **28**: 376–380; 1973.
12. Hansen, M. *Constitution of Binary Alloys.* 2nd ed. New York: McGraw-Hill; 1958; p. 620.
13. Morikawa, K.; Benedict, W.S.; and Taylor, H.S. "The activation of specific bonds in complex molecules at catalytic surfaces. II. The carbon–hydrogen and carbon–carbon bonds in ethane and ethane-*d*." *J. Amer. Chem. Soc.* **58**: 1795–1800; 1936.
14. Taylor, E.H. and Taylor, H.S. "The hydrogenation of ethane on cobalt catalysts." *J. Amer. Chem. Soc.* **61**: 503–509; 1939.
15. Cimino, A.; Boudart, M.; and Taylor, H.S. "Ethane hydrogenation-cracking on iron catalysts with and without alkali." *J. Phys. Chem.* **58**: 796–800; 1954.
16. Sinfelt, J.H. "Hydrogenolysis of ethane over supported platinum." *J. Phys. Chem.* **68**: 344–346; 1964.
17. Sinfelt, J.H.; Taylor, W.F.; and Yates, D.J.C. "Catalysis over supported metals. III. Comparison of metals of known surface area for ethane hydrogenolysis." *J. Phys. Chem.* **69**: 95–101; 1965.
18. Sinfelt, J.H. and Yates, D.J.C. "Catalytic hydrogenolysis of ethane over the noble metals of Group VIII." *J. Catal.* **8**: 82–90; 1967.
19. Sinfelt, J.H. and Yates, D.J.C. "Studies of ethane hydrogenolysis over Group VIII metals: supported osmium and iron." *J. Catal.* **10**: 362–367; 1968.
20. Yates, D.J.C. and Sinfelt, J.H. "An investigation of the dispersion and catalytic properties of supported rhenium." *J. Catal.* **14**: 182–186; 1969.
21. Sinfelt, J.H. "Kinetics of ethane hydrogenolysis." *J. Catal.* **27**: 468–471; 1972.
22. Sinfelt, J.H. "Specificity in catalytic hydrogenolysis by metals." *Advan. Catal.* **23**: 91–119; 1973.
23. Sinfelt, J.H. "Catalysis by metals: the P.H. Emmett Award address." *Catal. Rev.* **9***(1)*: 147–168; 1974.
24. Cotton, F.A. and Wilkinson, G. *Advanced Inorganic Chemistry.* 1st ed. New York: Interscience; 1962; p. 661; 760.
25. Carter, J.L.; Cusumano, J.A.; and Sinfelt, J.H. "Hydrogenolysis of *n*-heptane over unsupported metals." *J. Catal.* **20**: 223–229; 1971.
26. Sinfelt, J.H. "Heterogeneous catalysis by metals." *Progr. Solid State Chem.* **10** *(2)*: 55–69; 1975.
27. Pauling, L. "A resonating-valence-bond theory of metals and intermetallic compounds." *Proc. Roy. Soc. (London).* **A196**: 343–362; 1949.
28. Lam, Y.L. and Sinfelt, J.H. "Cyclohexane conversion on ruthenium catalysts of widely varying dispersion." *J. Catal.* **42**: 319–322; 1976.
29. Boudart, M. "Catalysis by supported metals." *Advan. Catal.* **20**: 153–166; 1969.

30. Schuit, G.C.A. and van Reijen, L.L. "The structure and activity of metal-on-silica catalysts." *Advan. Catal.* **10**: 242–317; 1958.
31. Cusumano, J.A.; Dembinski, G.W.; and Sinfelt, J.H. "Chemisorption and catalytic properties of supported platinum." *J. Catal.* **5**: 471–475; 1966.
32. Taylor, W.F. and Staffin, H.K. "Catalysis over supported nickel." *Trans. Faraday Soc.* **63**: 2309–2315; 1967.
33. Dixon, G.M. and Singh, K. "Catalysis over coprecipitated nickel–alumina." *Trans. Faraday Soc.* **65**: 1128–1137; 1969.
34. Aben, P.C.; van der Eijk, H.; and Oelderik, J.M. "The characterization of the metal surface of alumina-supported platinum catalysts by temperature-programmed desorption of chemisorbed hydrogen." In: Hightower, J.W., ed. *Proceedings of the Fifth International Congress on Catalysis.* Vol. 1. Amsterdam: North-Holland; 1973; p. 717–726.
35. Sinfelt, J.H. "Catalysis by alloys and bimetallic clusters." *Acc. Chem. Res.* **10**: 15–20;1977.
36. Anderson, J.R. and Baker, B.G. "Hydrocracking of neopentane and neohexane over evaporated metal films." *Nature (London).* **187**: 937–938; 1960.
37. Anderson, J.R.; MacDonald, R.J.; and Shimoyama, Y. "Relation between catalytic properties and structure of metal films. II. Skeletal reactions of some C_6 alkanes." *J. Catal.* **20**: 147–162; 1971.
38. Coles, B.R. "The lattice spacings of nickel–copper and palladium–silver alloys." *J. Inst. Metals.* **84**: 346–348; 1956.
39. Ahern, S.A.; Martin, M.J.C.; and Sucksmith, W. "The spontaneous magnetization of nickel–copper alloys." *Proc. Roy. Soc. (London).* **A248**: 145–152; 1958.
40. van der Plank, P. and Sachtler, W.M.H. "Surface composition of equilibrated copper–nickel alloy films." *J. Catal.* **7**: 300–303; 1967.
41. Cadenhead, D.A. and Wagner, N.J. "Low-temperature hydrogen adsorption on copper–nickel alloys." *J. Phys. Chem.* **72**: 2775–2781; 1968.
42. Helms, C.R. "Observation of the segregation of Cu to the surface of a clean, annealed, 50% Cu–50% Ni alloy by Auger electron spectroscopy." *J. Catal.* **36**: 114–117; 1975.
43. Williams, F.L. and Boudart, M. "Surface composition of nickel–gold alloys." *J. Catal.* **30**: 438–443; 1973.
44. Bouwman, R.; Lippits, G.J.M.; and Sachtler, W.M.H. "Photoelectric investigation of the surface composition of equilibrated Ag–Pd alloys in ultrahigh vacuum and in the presence of CO." *J. Catal.* **25**: 350–361; 1972.
45. Balandin, A.A. "The nature of active centers and the kinetics of catalytic dehydrogenation." *Advan. Catal.* **10**: 96–129; 1958.

46. Dowden, D.A. "Electronic structure and ensembles in chemisorption and catalysis by binary alloys." In: Hightower, J.W., ed. *Proceedings of the Fifth International Congress on Catalysis.* Vol. 1. Amsterdam: North-Holland; 1973; p. 621–631.

47. Roberti, A.; Ponec, V.; and Sachtler, W.M.H. "Reactions of methylcyclopentane on nickel and nickel–copper alloys." *J. Catal.* **28**: 381–390; 1973.

ABSTRACT

It was discovered that the ability of metals to form solid solutions (alloys) in the bulk is not necessary for a bimetallic system to be of interest as a catalyst. An example is the ruthenium–copper system, in which the two components are virtually completely immiscible in the bulk. This system exhibits an effect of the copper (in particular, selective inhibition of hydrocarbon hydrogenolysis) similar to that exhibited by the nickel–copper system, in which the components are completely miscible. Although ruthenium and copper do not form solid solutions in the bulk, they do exhibit a strong interaction at copper–ruthenium interfaces. The copper tends to cover the surface of the ruthenium, analogous to a chemisorbed layer. As a result, the copper has a marked effect on the chemisorption and catalytic properties of the ruthenium.

Chapter 3

BIMETALLIC SYSTEMS OF IMMISCIBLE COMPONENTS

The nickel–copper alloys discussed in the previous chapter were prepared under conditions of complete miscibility of the two components. In our exploratory studies on the chemisorption and catalytic properties of other bimetallic systems comprising a Group VIII and a Group IB metal, an interesting discovery was made: Systems which exhibit evidence of marked interaction between the components are not limited to combinations of metallic elements that form bulk alloys *(1–5)*.

Any notion that the components of a bimetallic catalyst should form a complete series of solid solutions in the bulk, or even that they should be moderately miscible, has been found to be much too restrictive. Indeed, pairs of metallic elements that are completely immiscible in the bulk, for example, ruthenium–copper and osmium–copper *(6),* may form bimetallic aggregates whose surface properties reveal extensive interaction between the two elements *(2,4)*.

The surface properties of unsupported ruthenium–copper aggregates are considered in this chapter. In a subsequent chapter on bimetallic cluster catalysts, the properties of supported ruthenium–copper and osmium–copper catalysts are considered in detail.

3.1 PREPARATION OF RUTHENIUM–COPPER AGGREGATES

The crystal structures of metallic ruthenium and copper are different, ruthenium having a hexagonal close-packed structure and copper a face-centered cubic structure *(7)*. Although the ruthenium–copper system can hardly be considered one which forms alloys, bimetallic ruthenium–copper aggregates can be prepared that are similar in their catalytic behavior to alloys such as nickel–copper *(3,4,8)*.

In one method of preparation, designated the *coprecipitation method,* an aqueous solution of ruthenium trichloride and copper nitrate is contacted with an aqueous solution of ammonia and hydrazine *(3).* The precipitate formed is dried and subsequently reduced with a dilute hydrogen stream (5 mole% hydrogen in helium). The reduction is conducted in stages, beginning with a 1-hour treatment of the material in the dilute hydrogen stream at 150°C, and continuing with an overnight treatment at 400°C.

In another method, designated the *sequential precipitation method,* an aqueous solution of ruthenium trichloride is contacted with the ammoniacal hydrazine solution *(3).* The resulting precipitate is filtered out of solution and reslurried in water, after which a solution of copper nitrate is added to the slurry. On subsequent addition of ammoniacal hydrazine solution, further precipitation occurs in the presence of the original precipitate. The total precipitate is then dried and reduced in the same manner as the coprecipitated preparations.

After reduction, the ruthenium–copper aggregates prepared by either method are cooled to room temperature in a stream of helium. They are then passivated by gradual admission of air to the helium and stored in closed containers until needed. In general, after these materials are charged to various kinds of experimental apparatus, they are again contacted with hydrogen at elevated temperature as a further step in preparing them for measurements of interest.

If portions of the materials are exposed to hydrogen at temperatures (500–600°C) higher than the 400°C employed in the initial preparation, the properties of the final ruthenium–copper aggregates are changed substantially, despite the fact that the total surface areas are not very different. The ruthenium–copper aggregates prepared by these procedures have surface areas in the approximate range of 5–7 m^2/g. In such aggregates the surface atoms constitute approximately 1% of the total metal atoms.

3.2 CHARACTERIZATION OF RUTHENIUM–COPPER AGGREGATES

Ruthenium–copper aggregates of the type described have been studied with chemical and physical probes. Chemical probes that have been very informative include hydrogen chemisorption and the hydrogenolysis of ethane to methane. Physical probes useful in these characterizations include X-ray diffraction and electron spectroscopy.

For highly dispersed ruthenium–copper catalysts (bimetallic clusters), to be discussed in the next chapter, useful information has also been obtained from high resolution electron microscopy and from analysis of extended X-ray absorption fine structure (EXAFS).

3.2.1 Characterization by Chemical Probes

In a typical hydrogen adsorption experiment, ruthenium–copper aggregates are first contacted with flowing hydrogen in the adsorption cell at 400°C to ensure thorough reduction. The cell is then evacuated to a pressure of approximately 10^{-6} torr and cooled to room temperature for adsorption measurements. Isotherms for total hydrogen adsorption and for weakly adsorbed hydrogen are then determined in the manner described for nickel–copper catalysts in Chapter 2.

Typical isotherms for ruthenium–copper aggregates containing 5 at.% copper are shown in Figure 3.1. The dashed line represents strongly adsorbed

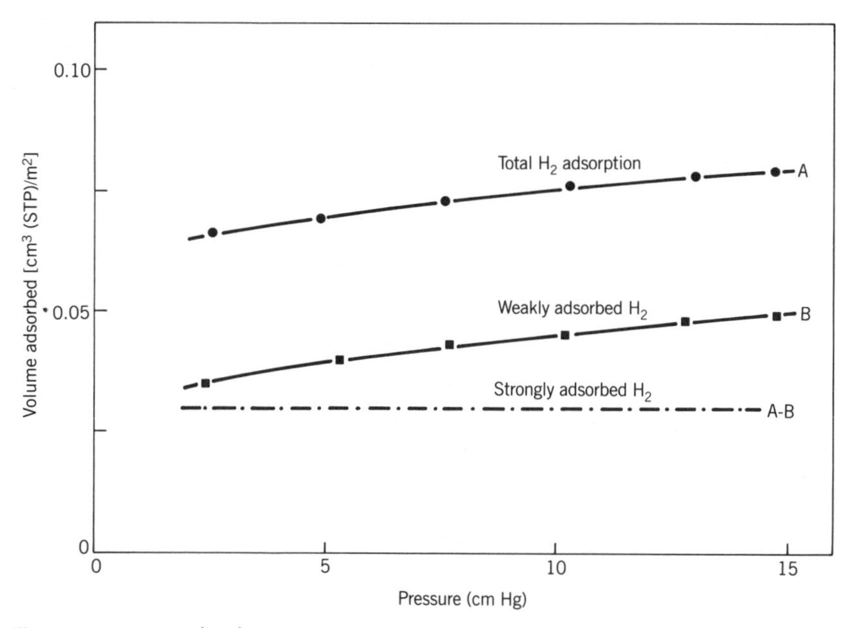

Figure 3.1 Typical adsorption isotherms at room temperature for hydrogen on ruthenium–copper aggregates containing 5 at.% copper *(3).* (Reprinted with permission from Academic Press, Inc.)

hydrogen, that is, the difference between the total hydrogen adsorption and the weakly adsorbed hydrogen.

In the upper field of Figure 3.2 the hydrogen chemisorption data refer to strongly adsorbed hydrogen. The volume adsorbed per square meter of surface is shown for ruthenium–copper aggregates of varying copper content.

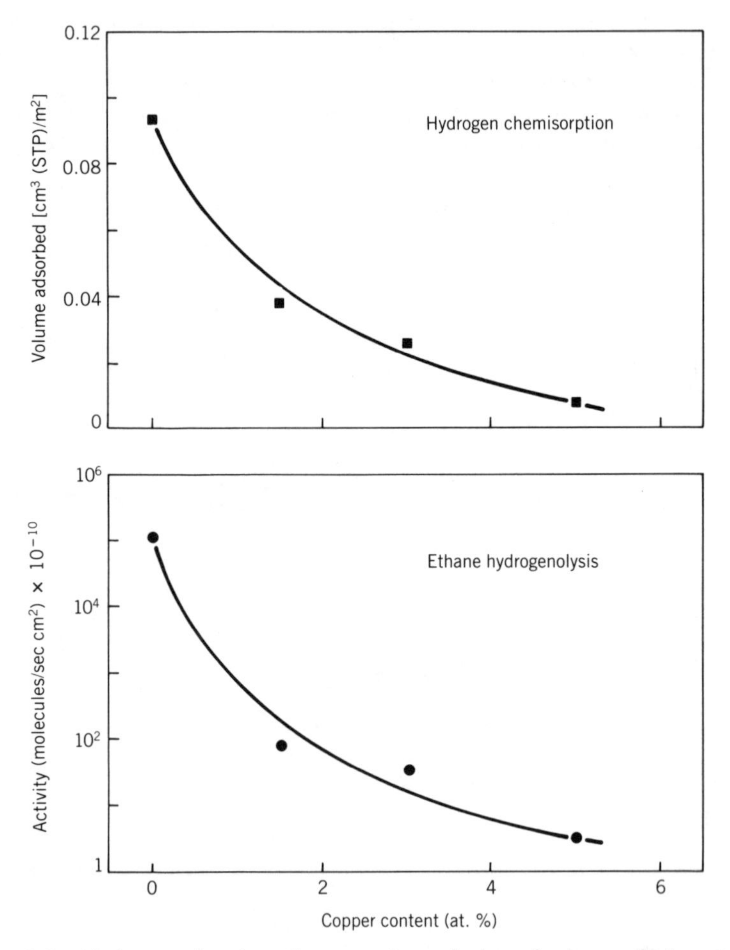

Figure 3.2 Hydrogen chemisorption capacity and ethane hydrogenolysis activity of ruthenium–copper aggregates as a function of copper content *(3)*. (Reprinted with permission from Academic Press, Inc.)

The materials were all heated in hydrogen at 500°C in their preparation *(3)*. The presence of copper with the ruthenium markedly decreases the amount of strongly adsorbed hydrogen.

The copper also strongly suppresses the catalytic activity of ruthenium for the hydrogenolysis of ethane to methane, as shown by the data in the lower field of Figure 3.2. The ethane hydrogenolysis activities are reaction rates measured at 245°C and ethane and hydrogen partial pressures of 0.030 and 0.20 atm, respectively.

The incorporation of only 1.5 at.% copper decreases the amount of strongly chemisorbed hydrogen by about 60% and lowers the activity for ethane hydrogenolysis by three orders of magnitude. With further increases in the amount of copper, the hydrogen chemisorption capacity and ethane hydrogenolysis activity continue to decrease markedly.

Thus although ruthenium and copper are immiscible in the bulk, ruthenium–copper aggregates can be prepared that have surface properties very different from those of pure ruthenium. The ruthenium–copper aggregates exhibit chemisorption and catalytic properties which would not be expected for simple physical mixtures of ruthenium and copper granules. The presence of the copper clearly has a marked effect on surface processes occurring on ruthenium. On the basis of the chemisorption and catalytic results, we conclude that the copper tends to cover the ruthenium surface. We thus adopt the view that copper is chemisorbed on the ruthenium.

The nature of the copper on the ruthenium surface is of interest. Monolayer coverage of ruthenium by copper would require the presence of approximately 1.5 at.% copper in the ruthenium–copper aggregates for which data are given in Figure 3.2. The highest copper content in Figure 3.2 thus corresponds to about three monolayers. The temperature of treatment of ruthenium–copper aggregates in hydrogen prior to their use in chemisorption and catalysis experiments has marked effects on the results. Such effects are not observed with pure ruthenium.

In Figure 3.3, data on hydrogen chemisorption (strongly adsorbed hydrogen) and ethane hydrogenolysis are given as a function of hydrogen treatment temperature for ruthenium–copper aggregates containing 1.5 at.% copper. The conditions used in obtaining the ethane hydrogenolysis activities are identical to those employed in the determination of the activities shown in Figure 3.2. The surface area of the aggregates decreased only slightly (about 27%) as a result of an increase in temperature from 400 to 600°C. Nevertheless, the hydrogen chemisorption capacity decreased by a factor of five, and the ethane hydrogenolysis activity declined by three orders of magnitude. These results suggest that coverage of the ruthenium by the copper increases

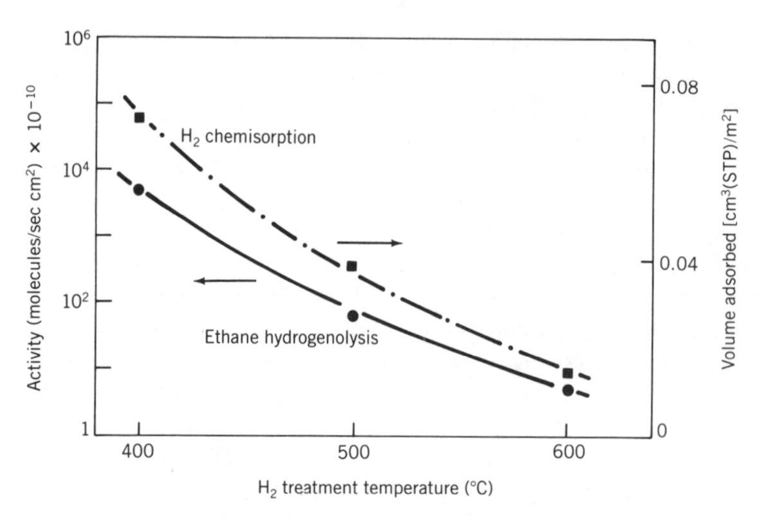

Figure 3.3 The effect of hydrogen treatment temperature on the hydrogen chemisorption capacity and ethane hydrogenolysis activity of ruthenium–copper aggregates containing 1.5 at.% copper *(3)*. (Reprinted with permission from Academic Press, Inc.)

with increasing temperature of hydrogen treatment in the range covered in Figure 3.3.

In aggregates treated at low temperatures, the copper at the surface is probably present in the form of three-dimensional clusters, leaving many small areas of pure ruthenium exposed at the surface. As the temperature is increased, the copper apparently spreads out over the surface, decreasing the amount of ruthenium exposed. The effect of temperature on the spreading of the copper is presumably associated with increased mobility of copper at the higher temperatures. The absolute temperature corresponding to 600°C is about two-thirds of the absolute melting temperature of copper. At such a temperature, the mobility of copper would be expected to be high.

There is a striking correlation between the ethane hydrogenolysis activity of ruthenium–copper aggregates and their capacity for strong hydrogen chemisorption, as illustrated in Figure 3.4 for a number of preparations of ruthenium–copper aggregates of varied composition. The hydrogenolysis activity is again the rate of hydrogenolysis at 245°C and ethane and hydrogen pressures of 0.030 and 0.20 atm, respectively. Strong hydrogen chemisorption is defined as in Figure 3.1. It has been concluded that the intermediate

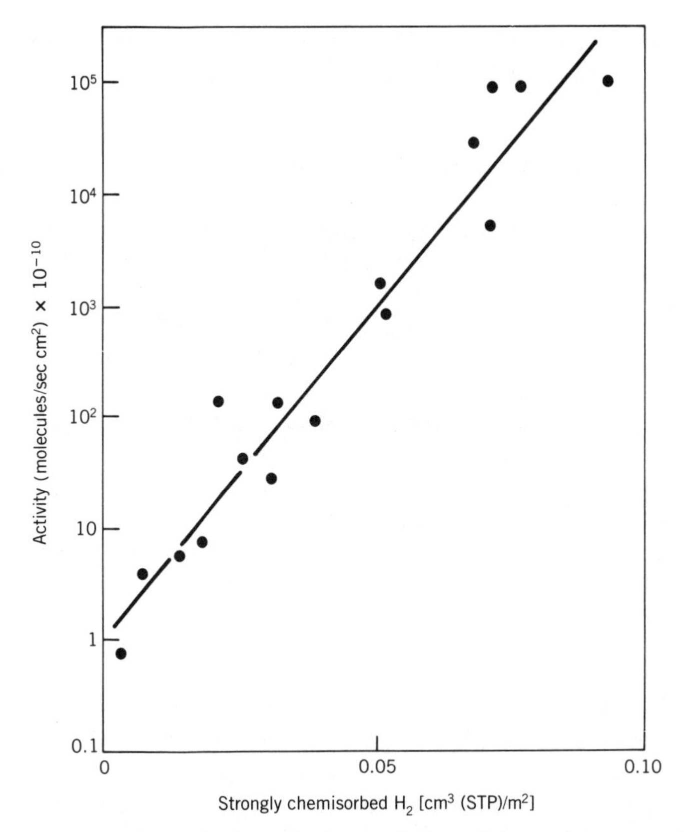

Figure 3.4 Correlation of ethane hydrogenolysis activity and amount of strongly chemisorbed hydrogen for ruthenium–copper aggregates *(3)*. (Reprinted with permission from Academic Press, Inc.)

in ethane hydrogenolysis is a dehydrogenated, dicarbon species that is bonded to more than one surface metal atom *(9–11)*. We conclude also that the rupture of the carbon–carbon bond in such an intermediate is facilitated by strong binding of the intermediate to the surface. Since it is reasonable for strong hydrocarbon chemisorption to correlate with strong hydrogen chemisorption, the relation between hydrogenolysis activity and strong hydrogen chemisorption is understandable.

The exponential nature of the relation in Figure 3.4 is a clear indication that ethane hydrogenolysis activity is much more sensitive to variations in the ruthenium–copper aggregates than is the capacity for strong hydrogen chemisorption. This relation suggests that the mode of chemisorption of ethane required for the hydrogenolysis reaction is more demanding than strong hydrogen chemisorption with regard to the nature of the surface site required. Thus the chemisorption of ethane in hydrogenolysis may require a site comprising a larger number of active metal atoms than is the case for strong hydrogen chemisorption. This could account for the result that a given amount of copper, particularly at the lowest copper levels, has a much greater effect on ethane hydrogenolysis activity than it does on the amount of strongly chemisorbed hydrogen. This type of explanation is appealing because of its simplicity. However, it is difficult to eliminate completely the possibility of a continuous decline in the intrinsic hydrogenolysis activity of structurally suitable sites as copper covers a progressively larger fraction of the surface.

In addition to hydrogen chemisorption and ethane hydrogenolysis, the reactions of cyclohexane provide a useful chemical probe for investigating ruthenium–copper aggregates. On pure ruthenium, two reactions of cyclohexane are readily observed: dehydrogenation to benzene and hydrogenolysis to lower carbon number alkanes. The product of the latter reaction is predominantly methane, even at very low conversions.

As shown in Figure 3.5, addition of copper to ruthenium decreases hydrogenolysis activity markedly but has a much smaller effect on dehydrogenation activity. The presence of copper thus improves the selectivity of conversion of cyclohexane to benzene. The data were obtained at 316°C with cyclohexane and hydrogen partial pressures of 0.17 and 0.83 atm, respectively. The ruthenium–copper aggregates were heated to 400°C in hydrogen in their preparation.

The different effects of copper on dehydrogenation and hydrogenolysis activities may be rationalized by arguments similar to those presented for nickel–copper alloy catalysts in Chapter 2. It is likely that different chemisorbed intermediates are involved in the dehydrogenation and hydrogenolysis reactions. The chemisorbed intermediate in hydrogenolysis is probably a hydrogen-deficient surface residue which forms a number of bonds with surface metal atoms. The probability of finding a suitable array, or multiplet *(12),* of active metal atoms to accommodate such an intermediate is greatly decreased when inactive copper is dispersed on the surface. By contrast, the chemisorbed intermediate in dehydrogenation presumably does not require a site consisting of a number of active metal atoms and is therefore relatively less sensitive to coverage of the surface with inactive copper.

The excellent maintenance of cyclohexane dehydrogenation activity as copper is incorporated with ruthenium, apart from illustrating the possibility

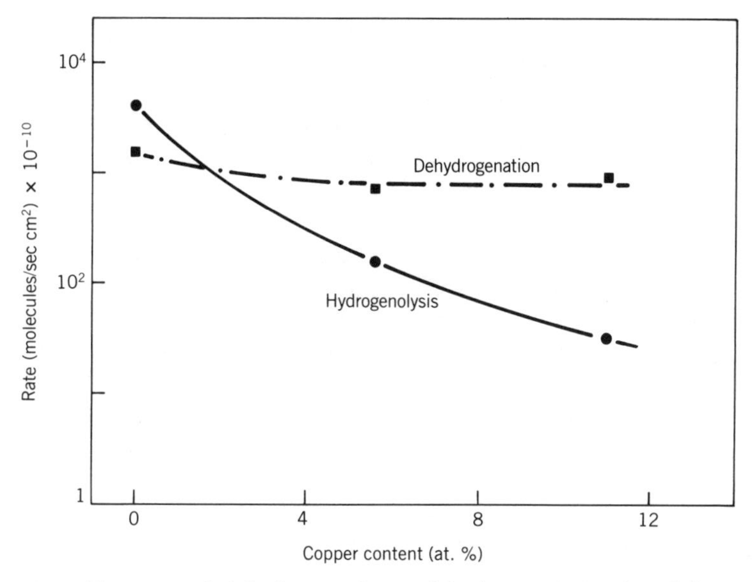

Figure 3.5 The rates of dehydrogenation and hydrogenolysis of cyclohexane on ruthenium–copper aggregates as a function of composition *(3)*. (Reprinted with permission from Academic Press, Inc.)

that this type of activity does not depend on the existence of multiplets of the active metal atoms, suggests the possibility that copper dispersed on the surface of ruthenium is itself active for the reaction. Sufficient information is not available for a definitive evaluation of this suggestion, but the possibility cannot be dismissed.

3.2.2 Characterization by Physical Probes

As noted in the previous section, monolayer coverage of ruthenium by copper in ruthenium–copper aggregates with total surface areas in the approximate range of 5–7 m^2/g is possible for a copper content in the vicinity of 1.5 at.%, if all the copper is present at the surface. X-ray diffraction patterns on ruthenium–copper aggregates of this type exhibit lines for ruthenium but not for copper. Such observations are consistent with a model of an aggregate in which copper is present as a thin layer or as small clusters at the surface of a crystallite composed essentially of pure ruthenium.

Additional evidence to support this model is provided by studies employing ESCA (electron spectroscopy for chemical analysis). In ESCA measurements, electrons are emitted from a substance as a result of irradiation by X-rays *(13).* In order for an electron to be emitted from an atom by an incident X-ray photon, the energy of the X-ray photon must be at least as high as the binding energy of the electron within the atom. The energy of the X-ray photon in excess of the binding energy is imparted to the emitted electron in the form of kinetic energy.

Since an atom possesses electrons with various binding energies corresponding to the different electron energy levels, irradiation by X-ray photons of a particular energy will result in the emission of electrons with a definite set of kinetic energies corresponding to lines in an ESCA spectrum. The intensity of a spectral line in ESCA is a measure of the number of emitted electrons with a particular kinetic energy. From the kinetic energy of the emitted electrons associated with a particular line and the known energy of the incident X-ray photons, one can determine the binding energy of an electron in a particular level in the emitting atom.

As the electron binding energy increases from one level to another, the kinetic energy of the emitted electrons decreases. Since an electron binding energy is a characteristic feature of the atoms of a given element, the determination of binding energies by ESCA makes it possible to identify the kinds of elements present in a material. Data on intensities of the lines provide information on the amounts of elements present. Typical ESCA lines for electron emissions from ruthenium atoms in pure ruthenium and in a ruthenium–copper catalyst are shown in Figure 3.6. Lines for emissions from copper atoms in pure copper and in a ruthenium–copper catalyst are shown in Figure 3.7 *(14).*

A feature of particular interest in ESCA is its surface sensitivity. Electrons emitted by atoms in subsurface layers of a material are subject to inelastic scattering processes. These processes decrease the probability that the electrons will emerge from the material with the expected kinetic energy. The deeper the atoms are located in the material, the lower this probability becomes. For atoms at a given depth, the probability of an electron escaping from a material with its kinetic energy unchanged depends on the magnitude of the kinetic energy. In other words, the mean free path of the electrons for inelastic scattering is a function of the kinetic energy, as illustrated in Figure 3.8. The band in the figure was derived from data on many different systems and is useful for making estimates of electron mean free paths *(14–16).*

In ESCA studies of ruthenium–copper aggregates we have used the ratio of the intensity of the Cu $2p_{3/2}$ line to that of the Ru $3d_{5/2}$ line as a measure of the degree of coverage of ruthenium by copper *(14).* The kinetic energies

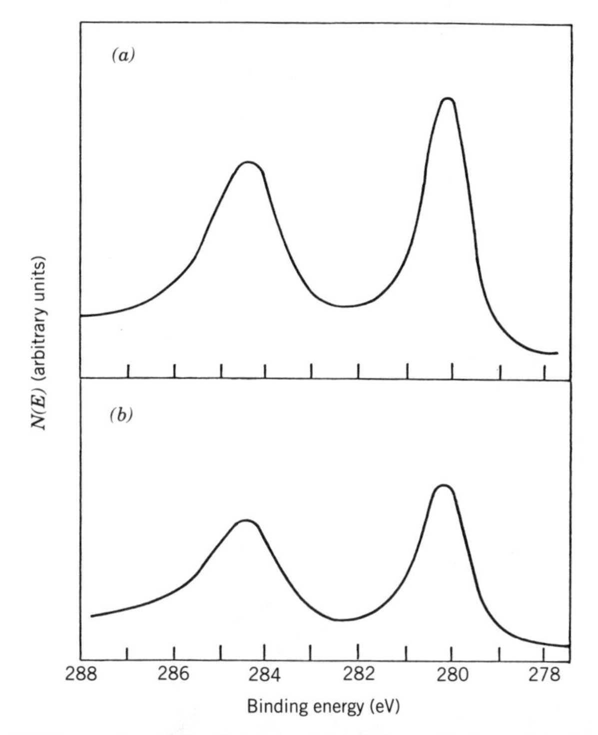

Figure 3.6 ESCA spectra of Ru $3d_{3/2}$ and Ru $3d_{5/2}$ emissions (binding energies of 284.4 and 280.2 eV, respectively) from (*a*) pure Ru metal, and (*b*) Ru–Cu aggregates containing 5 at.% Cu after heating in vacuum at 300°C for 30 min *(14)*. (Reprinted with permission from North-Holland Publishing Company.)

of the emitted electrons corresponding to the copper and ruthenium lines are approximately 550 and 1200 eV, respectively. In a simple situation in which a monolayer of copper of thickness d covers ruthenium, it can be shown that R, the ratio of emitted electrons arising from the copper layer to those arising from the ruthenium, is given approximately by the expression

$$R = R_0[1 - \exp(-d/L_{Cu})] \exp(d/L_{Ru}) \qquad (3.1)$$

where R_0 is the ratio of the intensity of the $2p_{3/2}$ line of pure metallic copper to that of the $3d_{5/2}$ line of pure metallic ruthenium. The quantities L_{Cu} and L_{Ru} are the effective mean free paths of the electrons emitted from copper

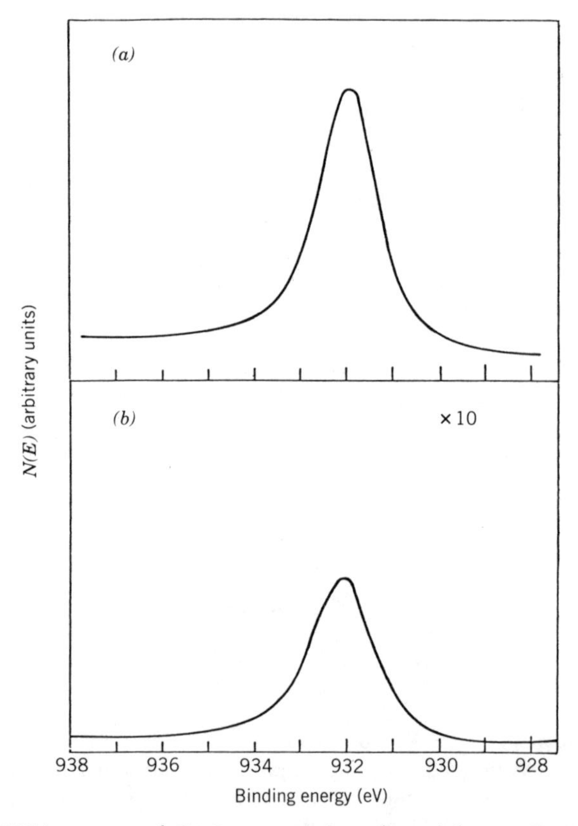

Figure 3.7 ESCA spectra of Cu $2p_{3/2}$ emissions from (*a*) pure Cu metal, and (*b*) Ru–Cu aggregates containing 5 at.% Cu after heating in vacuum at 300°C for 30 min *(14)*. (Reprinted with permission from North-Holland Publishing Company.)

and ruthenium. They are equal to the actual mean free paths multiplied by the cosine of the average electron takeoff angle, the angle being measured with respect to the surface normal. In the experiments to be considered here, the average takeoff angle was approximately 45°.

In estimates of electron mean free paths from Figure 3.8, there is uncertainty indicated by the width of the band. As a result, the value of R/R_0 estimated from Eq. 3.1 ranges from 0.25 to 0.50 for the electron energies of interest, depending on the values taken for electron mean free paths from the band in the figure. Since a value of R_0 equal to 1.0 ± 0.1 is obtained from experi-

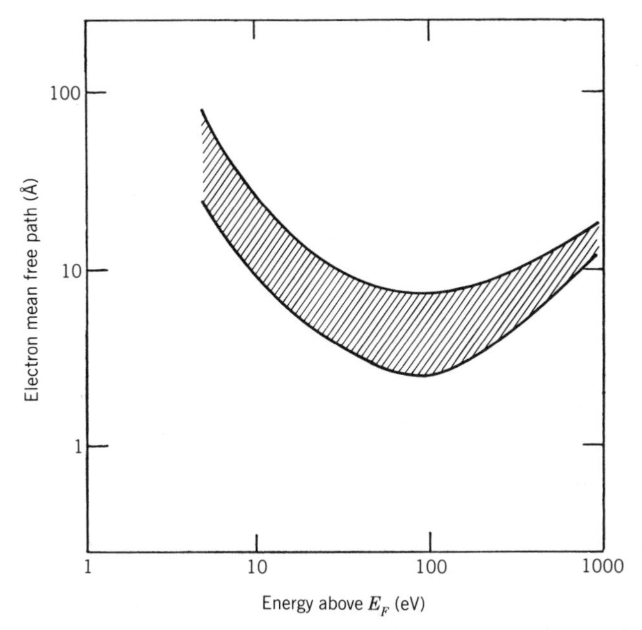

Figure 3.8 Dependence of electron mean free path on the kinetic energy of the electrons *(14–16)*. (Reprinted with permission from North-Holland Publishing Company.)

ments on pure copper and ruthenium crystals, the value of R also ranges from 0.25 to 0.50. The large uncertainty in the value precludes the use of the ESCA data for obtaining precise quantitative results on the degree of coverage of ruthenium by copper. Nevertheless, experimental values of the ratio I_{Cu}/I_{Ru} of the intensity of electron emission from the copper to that from the ruthenium are useful in assessing trends of variation in the degree of coverage of ruthenium by copper.

In Figure 3.9 the ratio of the intensity of the copper $2p_{3/2}$ peak to that of the ruthenium $3d_{5/2}$ peak is shown as a function of copper content for ruthenium–copper aggregates heated in hydrogen in situ at 500°C in the ESCA chamber *(14, 17)*. The chamber was evacuated prior to the intensity measurements. The dashed line shows the relationship expected if the ruthenium and copper were uniformly distributed throughout the samples, as determined from ESCA data on pure ruthenium and copper. The solid curve shows the data actually obtained. Intensity ratios much higher than those corresponding to the dashed line are observed, indicating that the external boundary sur-

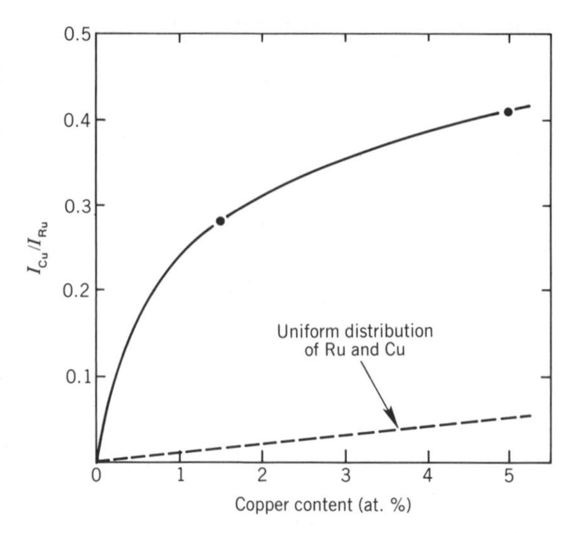

Figure 3.9 Electron spectroscopy (ESCA) data showing the ratio I_{Cu}/I_{Ru} of the intensity of electron emission from the copper $2p_{3/2}$ level to that from the ruthenium $3d_{5/2}$ level for ruthenium–copper aggregates as a function of copper content *(14,17)*. (Dashed line represents the ratio expected for a uniform distribution of copper with ruthenium.)

faces of the ESCA samples have higher Cu/Ru atomic ratios than correspond to the overall sample compositions. These results are consistent with a model in which copper tends to cover the ruthenium in the bimetallic aggregates.

In Figure 3.10 the same ratio of intensities is shown as a function of the temperature of hydrogen treatment for ruthenium–copper aggregates containing 1.5 at.% copper *(14).* The ratio increases approximately twofold, from 0.2 to 0.4, when the hydrogen treatment temperature is increased from 400 to 600°C. The increase in this range of temperature indicates an increase in the extent of coverage of the ruthenium by the copper. The results are consistent with our earlier conclusions based on hydrogen chemisorption and ethane hydrogenolysis studies, in which it was concluded that the increased mobility of the copper at the higher temperature permitted it to spread over the surface.

It is of interest to relate ESCA data on ruthenium–copper aggregates to data on hydrogen chemisorption on these materials, since changes in the degree of coverage of ruthenium by copper affect both the Cu/Ru intensity ratio and hydrogen chemisorption capacity. Such a relationship is shown in Figure 3.11

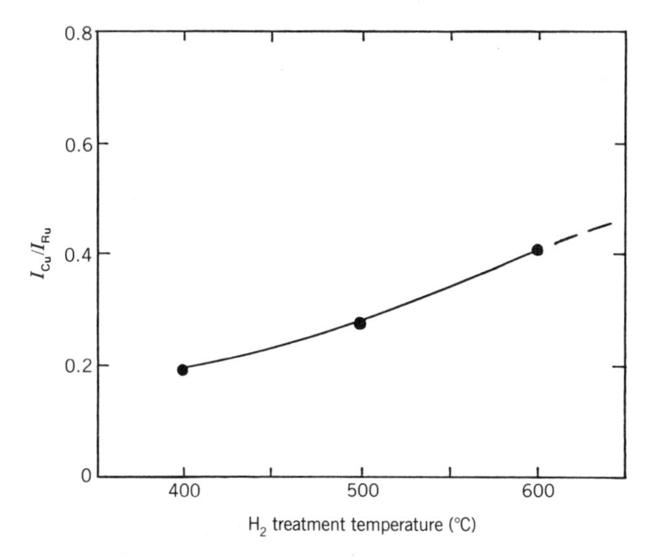

Figure 3.10 Effect of the temperature of hydrogen pretreatment of Ru–Cu aggregates containing 1.5 at.% Cu on the ratio I_{Cu}/I_{Ru} of the intensity of electron emission from the copper $2p_{3/2}$ level to that from the ruthenium $3d_{5/2}$ level *(14)*. (Reprinted with permission from North–Holland Publishing Company.)

for ruthenium–copper aggregates heated in hydrogen at 400 to 600°C and evacuated at elevated temperature prior to the ESCA and hydrogen chemisorption measurements *(14)*. The amount of hydrogen chemisorbed at room temperature, expressed as molecules of hydrogen adsorbed per square centimeter of surface, is shown as a function of the Cu/Ru intensity ratio obtained in the ESCA measurements *(14)*. As noted earlier, the intensity data are for the copper $2p_{3/2}$ and ruthenium $3d_{5/2}$ peaks.

The upper part of Figure 3.11 shows data on the total chemisorption of hydrogen (at 100-torr equilibrium pressure). The lower part shows data on the strongly chemisorbed fraction, that is, the amount which cannot be removed by evacuation to 10^{-6} torr at room temperature. Both total and strong chemisorption decrease as the Cu/Ru intensity ratio increases, but the percentage decrease is greater for the strong chemisorption.

Pure copper exhibits no strong chemisorption of hydrogen. Data on strong hydrogen chemisorption on ruthenium–copper aggregates have been used as a measure of the amount of ruthenium surface not covered by copper *(3,4)*. As the coverage by copper increases, the amount of strongly chemi-

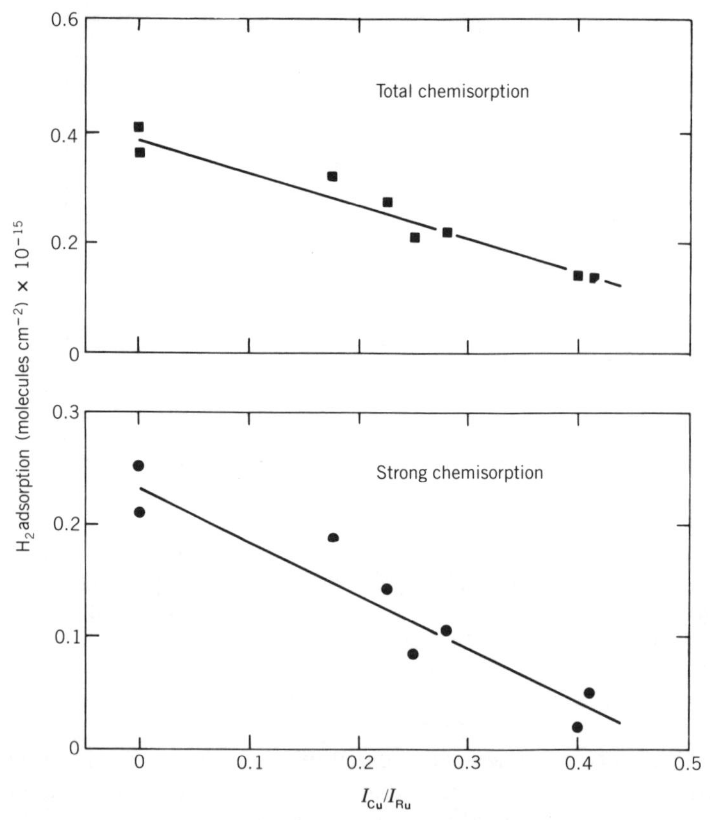

Figure 3.11 Correlation of hydrogen chemisorption capacity of ruthenium–copper aggregates with the ratio I_{Cu}/I_{Ru} of the intensity of electron emission from the copper $2p_{3/2}$ level to that from the ruthenium $3d_{5/2}$ level *(14)* (chemisorption data at room temperature). (Reprinted with permission from North-Holland Publishing Company.)

sorbed hydrogen decreases. While the total chemisorption also decreases, the effect is more complex, since there is weak chemisorption of hydrogen on both copper and ruthenium.

The data in the upper part of Figure 3.11 show a total hydrogen chemisorption of 0.4×10^{15} molecules cm^{-2} on the pure ruthenium aggregates, which corresponds to 0.8×10^{15} atoms cm^{-2}. This is somewhat lower than one might expect for monolayer coverage on ruthenium, perhaps 30% lower, and indicates either that such a monolayer is not quite attained at a pressure of 100 torr or that the ruthenium surface is partially contaminated.

3.3 RELATED STUDIES OF OTHER INVESTIGATORS

Recent work conducted in the laboratory of G. Ertl in Munich has extended the investigations on the ruthenium–copper system to include single crystal specimens *(18–20)*. The results of the work are in excellent accord with those obtained in our laboratory on unsupported ruthenium–copper aggregates and on supported ruthenium–copper clusters as well. Our work on supported bimetallic clusters of ruthenium and copper is discussed in detail in the following chapter.

The studies of Ertl and associates employed a surface consisting of the *(0001)* face of a ruthenium single crystal. Copper was deposited on the ruthenium surface by exposing the surface to a flux of copper atoms obtained by evaporation of a copper source. From low energy electron diffraction, Auger electron spectroscopy, thermal desorption, and work function measurements *(18)* Ertl and associates concluded that copper deposits on the ruthenium surface at 540°K in the form of a two-dimensional overlayer to coverages of 50 to 60%, beyond which there is a transition to a three-dimensional growth phase.

They further suggest that it may be possible to form a complete monolayer of copper on ruthenium by increasing the temperature at which the copper is deposited. This suggestion is consistent with our conclusion that the spreading of copper on ruthenium is aided by increasing the temperature of exposure of ruthenium–copper aggregates to hydrogen in their preparation *(3,14)*. The beneficial effect of the higher temperature is presumably the result of enhanced mobility of the copper.

From thermal desorption data, Ertl and associates found that the strength of bonding of a copper atom to the ruthenium surface is greater by several kilocalories per mole than its strength of bonding to copper *(18)*, which accounts for the tendency of copper to spread over the surface of ruthenium rather than to form three-dimensional copper crystallites. These investigators also observed that the deposition of copper on ruthenium decreased the amounts of strongly adsorbed hydrogen *(19)* and carbon monoxide *(20)* markedly, in further agreement with our early observations *(2,3)*.

The marked similarity between the results of the Ertl group on single crystals and the results from our laboratory on actual catalytic materials provides good evidence that studies on single crystals can provide information that is relevant to catalysts. Although this finding is not necessarily surprising, it is nevertheless reassuring to have convincing experimental results relating the two different approaches.

REFERENCES

1. Sinfelt, J.H.; Barnett, A.E.; and Carter, J.L., inventors; Exxon Research and Engineering Company, assignee. "Inhibition of hydrogenolysis." U.S. Patent 3,617,518. 4 pages. November 2, 1971.
2. Sinfelt, J.H. "Supported 'bimetallic cluster' catalysts." *J. Catal.* **29**: 308–315; 1973.
3. Sinfelt, J.H.; Lam, Y.L.; Cusumano, J.A.; and Barnett, A.E. "Nature of ruthenium–copper catalysts." *J. Catal.* **42**: 227–237; 1976.
4. Sinfelt, J.H. "Catalysis by alloys and bimetallic clusters." *Acc. Chem. Res.* **10:** 15–20; 1977.
5. Sinfelt, J.H. "Bimetallic clusters in catalysis." *The Chemist.* **59***(3)*: 8–12; 1982.
6. Hansen, M. *Constitution of Binary Alloys.* 2nd ed. New York: McGraw-Hill; 1958; p. 607, 620.
7. Cullity, B.D. *Elements of X-ray Diffraction.* Reading, MA: Addison-Wesley; 1956; p. 482, 484.
8. Sinfelt, J.H. "Heterogeneous catalysis: some recent developments." *Science.* **195**: 641–646; 1977.
9. Cimino, A.; Boudart, M.; and Taylor, H.S. "Ethane hydrogenation-cracking on iron catalysts with and without alkali." *J. Phys. Chem.* **58**: 796–800; 1954.
10. Sinfelt, J.H. "Kinetics of ethane hydrogenolysis." *J. Catal.* **27**: 468–471; 1972.
11. Sinfelt, J.H. "Specificity in catalytic hydrogenolysis by metals." *Advan. Catal.* **23**: 91–119; 1973.
12. Balandin, A.A. "The nature of active centers and the kinetics of catalytic dehydrogenation." *Advan. Catal.* **10**: 96–129; 1958.
13. Brundle, C.R. "The application of electron spectroscopy to surface studies." *J. Vacuum Sci. Technol.* **11***(1)*: 212–224; 1974.
14. Helms, C.R. and Sinfelt, J.H. "Electron spectroscopy (ESCA) studies of Ru–Cu catalysts." *Surface Sci.* **72**: 229–242; 1978.
15. Helms, C.R. and Yu, K.Y. "Determination of the surface composition of the Cu–Ni alloys for clean and adsorbate-covered surfaces." *J. Vacuum Sci. and Technol.* **12**: 276–278; 1975.
16. Lindau, I. and Spicer, W.E. "The probing depth in photoemission and Auger-electron spectroscopy." *J. Electron Spectrosc.* **3**: 409–413; 1974.
17. Sinfelt, J.H. "Structure of metal catalysts." *Rev. Mod. Phys.* **51**: 569–589; 1979.
18. Christmann, K.; Ertl, G.; and Shimizu, H. "Model studies on bimetallic Cu/Ru catalysts. I. Cu on Ru*(0001).*" *J. Catal.* **61**: 397–411; 1980.
19. Shimizu, H.; Christmann, K.; and Ertl, G. "Model studies on bimetallic Cu/Ru catalysts. II. Adsorption of hydrogen." *J. Catal.* **61**: 412–429; 1980.
20. Vickerman, J.C.; Christmann, K.; and Ertl, G. "Model studies on bimetallic Cu/Ru catalysts. III. Adsorption of carbon monoxide." *J. Catal.* **71**: 175–191; 1981.

ABSTRACT

For a bimetallic catalyst to be of practical interest, a high surface area is generally necessary. An effective approach to creating high surface area is the dispersal of the bimetallic entity on a carrier. The highly dispersed bimetallic species thus obtained are called *bimetallic clusters.* This term is used in preference to highly dispersed alloys, since the systems of interest are not limited to combinations of metallic elements that form alloys in the usual sense (i.e., solid solutions in the bulk). The bimetallic clusters of interest for catalysis are generally smaller than about 100 Å and are commonly in the 10- to 50-Å range. In some cases the clusters are so small that virtually every metal atom is a surface atom. Information on the structures of bimetallic clusters has been obtained with chemical probes, such as chemisorption and reaction rate measurements, and physical probes, such as X-ray diffraction, extended X-ray absorption fine structure (EXAFS), and Mössbauer effect spectroscopy.

Chapter 4

BIMETALLIC CLUSTER CATALYSTS

The term *bimetallic clusters* has been proposed by the author in referring to highly dispersed metallic entities comprising atoms of two metallic elements *(1,2)*. As a rough guideline, the clusters of interest have sizes smaller than about 100 Å and they are commonly in the size range of 10 to 50 Å. In some cases the clusters are so small that virtually every atom in the cluster is a surface atom. In general, the clusters are supported on a high surface area carrier such as silica or alumina, and the total metal content of the system may be as little as 1% or lower.

These novel systems introduce much flexibility into the design of catalysts. Virtually any property of a dispersed metal catalyst (including activity, selectivity, and stability) may be influenced by addition of a second metallic element to form bimetallic clusters. Systems of interest are not limited to combinations of metals that form bulk alloys.

Two types of bimetallic clusters have been investigated extensively by the author. One is a combination of a Group VIII and a Group IB metal, for example, ruthenium–copper and osmium–copper *(1)*. The other is a combination of two Group VIII metals, for example, platinum–iridium *(3,4)*. Another system which could conceivably be considered in the bimetallic cluster category is supported platinum–rhenium *(5,6)*, which is a combination of a Group VIII metal and a Group VIIA metal. Both platinum–iridium and platinum–rhenium catalysts are of interest in the reforming of petroleum fractions.

Supported bimetallic clusters can be prepared simply by contacting a suitable carrier such as silica or alumina with an aqueous solution of salts of the two metals of interest. The material is then dried and contacted with a stream of hydrogen at a temperature high enough to accomplish reduction of the metal precursors to the metallic state. This procedure results in the formation of very small metal crystallites or clusters dispersed on the surface of the carrier.

The nature of these metallic entities is a key question. At the outset we had the task of demonstrating that mixing of atoms of the two metallic elements occurred to give bimetallic clusters on the carrier rather than monometallic clusters of each. On purely statistical grounds it might be expected that the individual clusters would contain atoms of both metallic elements, especially if the two elements form solid solutions in the bulk. This expectation has been supported by experiment, even for cases in which the individual components exhibit extremely low miscibility in the bulk.

4.1 RUTHENIUM–COPPER AND OSMIUM–COPPER CLUSTERS

Ruthenium–copper and osmium–copper clusters are examples of bimetallic clusters in which one component is from Group VIII and the other from Group IB of the periodic table. These clusters are of particular interest because copper is virtually completely immiscible with either ruthenium or osmium in the bulk *(7)*. Copper has the face-centered cubic structure in the metallic state, whereas ruthenium and osmium both exhibit hexagonal close-packed structures *(8)*.

4.1.1 Characterization by Chemical Probes

Silica-supported ruthenium–copper and osmium–copper catalysts containing 1 wt% of either ruthenium or osmium and varying amounts of copper have been investigated *(1)*. The atomic ratio of copper to the Group VIII metal ranged from 0 to 1. The degree of metal dispersion was high in all of these catalysts, as evidenced by the fact that X-ray diffraction scans showed no lines due to the metals. Electron microscopy data indicated average diameters of 32 and 10 Å, respectively, for the metal clusters present in the ruthenium–copper and osmium–copper catalysts *(9)*.

Direct experimental verification of very highly dispersed bimetallic clusters is complicated by limitations in the ability of physical methods to obtain structural information on such systems. In such a system, however, a catalytic reaction can serve as a sensitive probe to obtain evidence of interaction between the atoms of the two metallic components. For supported bimetallic combinations of a Group VIII and a Group IB metal, the hydrogenolysis of ethane to methane is a useful reaction for this purpose. In the case of unsupported bimetallic systems of this type, as discussed previously, the interaction between the Group VIII metal and the Group IB metal results in a marked suppression of the hydrogenolysis activity of the former.

Interaction at a surface does not require that the components form alloys, as shown by results on the ruthenium–copper system. In applying ethane hydrogenolysis as a probe to establish interaction between copper and either ruthenium or osmium on a carrier, one simply looks for a marked suppression of hydrogenolysis activity of the Group VIII metal when copper is present. Data on hydrogenolysis activities of silica-supported ruthenium–copper and osmium–copper catalysts containing 1 wt% of either ruthenium or osmium and varying amounts of copper are shown in Figure 4.1. Activities were determined at 245°C and ethane and hydrogen partial pressures of 0.030 and 0.20 atm, respectively. Since incorporation of copper with the Group VIII metal decreases the hydrogenolysis activity markedly, the metal components clearly are not isolated from each other on the carrier. The data provide

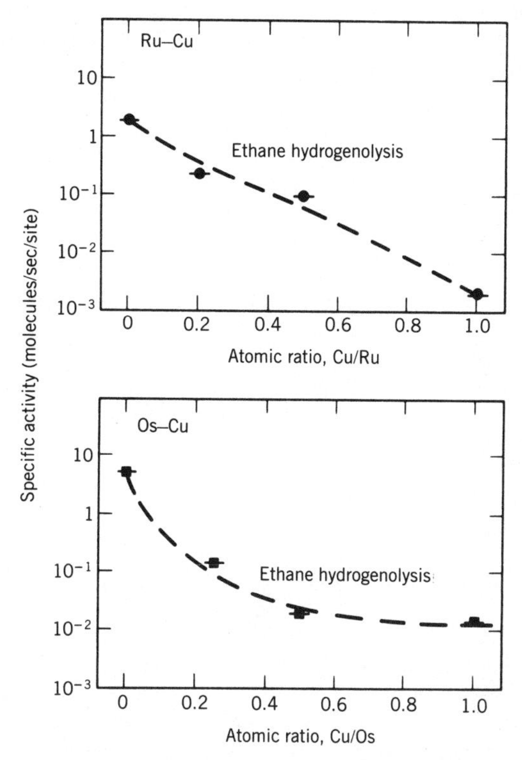

Figure 4.1 The specific activities of silica-supported ruthenium–copper and osmium–copper catalysts for ethane hydrogenolysis *(1,2)*. (Reprinted with permission from Academic Press, Inc.)

strong evidence of interaction between copper and the Group VIII metal and thus support the bimetallic cluster concept. It is particularly intriguing that such an effect is observed with supported ruthenium–copper and osmium–copper, in view of the extremely low miscibility of copper with ruthenium or osmium in the bulk state.

It is of interest to consider how the state of dispersion of ruthenium–copper catalysts affects the dependence of hydrogen chemisorption capacity and ethane hydrogenolysis activity on catalyst composition *(2,10,11)*. Comparisons of large ruthenium–copper aggregates with highly dispersed, silica-supported ruthenium–copper clusters are given in Figures 4.2 and 4.3. The dispersion, defined as the ratio of surface atoms to total metal atoms, is of the order of 0.01 for the large ruthenium–copper aggregates and 0.5 for the ruthenium–copper clusters. In addition to having an inhibiting effect on the hydrogenolysis activity of ruthenium, the presence of the copper also decreases the hydrogen chemisorption capacity. The atomic ratio of copper to ruthenium corresponding to a given degree of suppression of hydrogen chemisorption capacity, or to a given decrease in ethane hydrogenolysis ac-

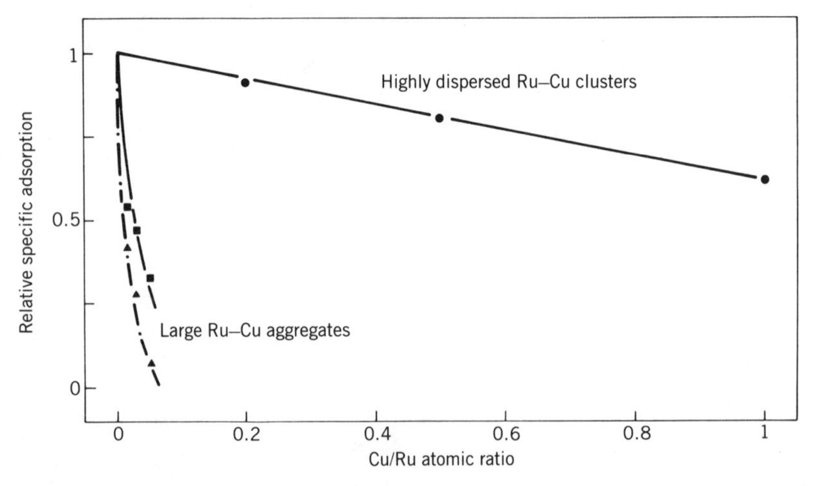

Figure 4.2 Influence of the state of dispersion of ruthenium–copper catalysts on the relationship between hydrogen chemisorption capacity and catalyst composition *(10, 11)*. (Square points for large ruthenium–copper aggregates represent total hydrogen chemisorption; triangular points represent strongly chemisorbed hydrogen, i.e., hydrogen which is not removed from the surface by evacuation of the adsorption cell at room temperature to a pressure of approximately 10^{-6} torr). (Reprinted with permission from Academic Press, Inc.)

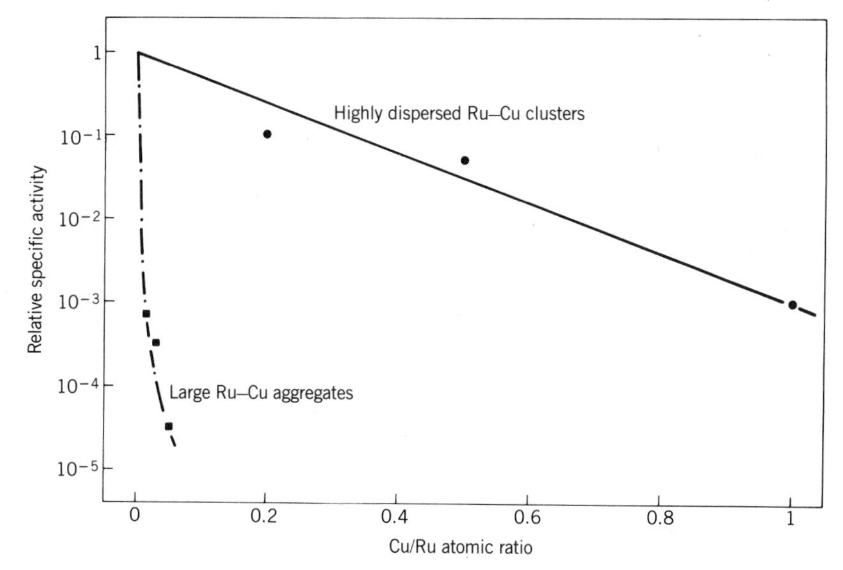

Figure 4.3 Influence of the state of dispersion of ruthenium–copper catalysts on the relationship between ethane hydrogenolysis activity and catalyst composition *(2,10, 11)*. (Reprinted with permission from Academic Press, Inc.)

tivity, is observed to be much lower for large ruthenium–copper aggregates than for highly dispersed ruthenium–copper clusters, as shown in Figures 4.2 and 4.3. The observed ratios differ by a factor approximately equal to the ratio of the metal dispersions. This indicates that the copper in a ruthenium–copper aggregate is confined to the surface, which is consistent with the extremely low miscibility of the two metals in the bulk state. If one envisions a series of ruthenium–copper aggregates of varying size, each containing a monolayer of copper on the surface, the atomic ratio of copper to ruthenium in an aggregate will increase with decreasing aggregate size. When the aggregate becomes sufficiently small, the number of atoms in the surface approaches the number in the interior, and the atomic ratio of copper to ruthenium approaches unity. The resulting ruthenium–copper entity may then be taken as a model of the bimetallic clusters in a highly dispersed ruthenium–copper catalyst containing equiatomic amounts of copper and ruthenium. From this simple consideration, it is readily seen how highly dispersed bimetallic clusters may have compositions far outside the range of those possible in a bulk solid solution of the two metals.

The selectivity aspect of catalysis with bimetallic systems is again illustrated by data on cyclohexane conversion on ruthenium–copper and osmium–copper cluster catalysts *(1, 12).* Two reactions are observed, dehydrogenation to benzene and hydrogenolysis to alkanes. As copper is incorporated with either ruthenium or osmium, the hydrogenolysis activity is strongly inhibited, but the dehydrogenation activity is relatively unaffected (Figure 4.4). The activities in Figure 4.4 were determined at 316°C and cyclohexane and hydrogen pressures of 0.17 and 0.83 atm, respectively. The selectivity, defined as the ratio of the dehydrogenation activity D to the hydrogenolysis activity H, improves markedly as a consequence (Figure 4.5). The behavior is similar to

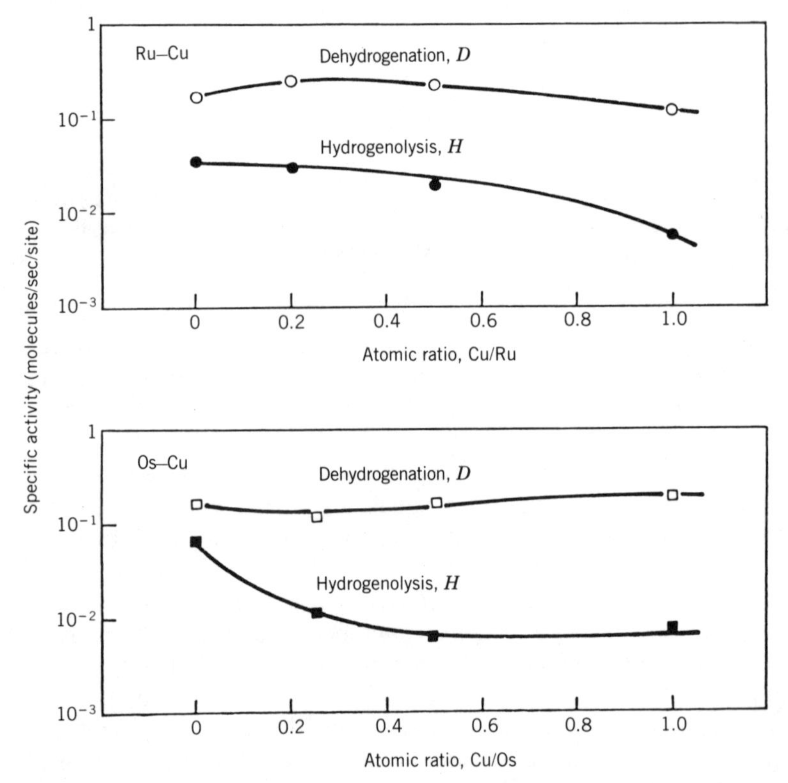

Figure 4.4 The specific activities of silica-supported ruthenium–copper and osmium–copper catalysts for the dehydrogenation and hydrogenolysis of cyclohexane, the former reaction yielding benzene and the latter alkanes *(1).* (Reprinted with permission from Academic Press, Inc.)

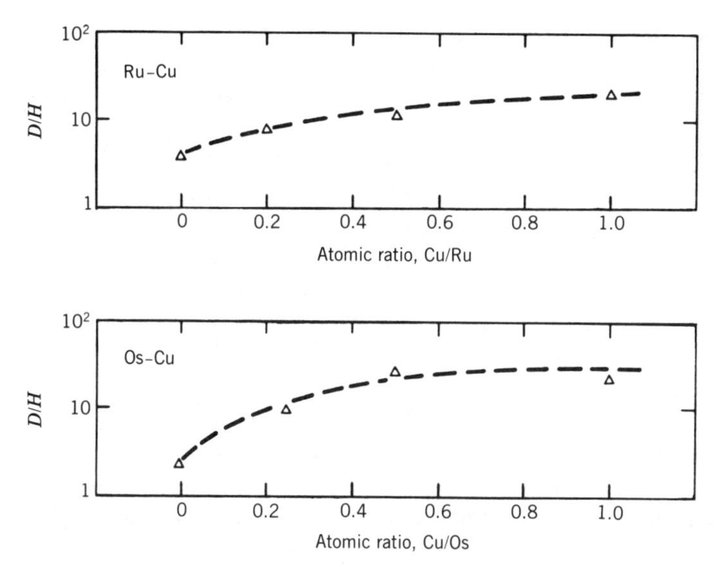

Figure 4.5 Selectivity of conversion of cyclohexane over silica-supported bimetallic clusters of ruthenium–copper and osmium–copper at 316°C, as represented by the ratio D/H *(1, 12)*. (D is rate of dehydrogenation of cyclohexane to benzene, and H is rate of hydrogenolysis to alkanes.) (Reprinted with permission from Academic Press, Inc.)

that described in Chapter 3 for unsupported ruthenium–copper aggregates, providing further evidence for the interaction between copper and the Group VIII metal on the carrier.

4.1.2 Characterization by Physical Probes

When the initial research on bimetallic clusters such as ruthenium–copper and osmium–copper was conducted, the characterization of the clusters was limited to methods involving chemical probes because of the difficulty of obtaining information with physical probes. In recent years, however, advances in X-ray absorption spectroscopy have changed the situation markedly. In particular, improvements in methods of obtaining extended X-ray absorption fine structure (EXAFS) data with the use of synchrotron radiation *(13)*, in conjunction with advances in methods of data analysis *(14)*, have made EXAFS a valuable tool for obtaining structural information on bimetallic clusters.

Description of EXAFS. In the absorption of X-rays by matter other than monatomic gases, a plot of absorption coefficient vs. X-ray energy exhibits an extended fine structure on the high energy side of an absorption edge. The fine structure constituting EXAFS consists of fluctuations in the absorption coefficient beginning at an energy of approximately 30 eV beyond the edge and extending over a range of 1000 to 1500 eV. From an analysis of EXAFS data, one can obtain information on the number and type of neighboring atoms about a given absorber atom, the interatomic distances, and the root mean square deviations of distances about the equilibrium or average values. By determining EXAFS for each element of interest in a complex material, one can obtain information on the environment of each type of atom present.

While the phenomenon of EXAFS has been known for a long time *(15–17),* its possibilities for investigating the structures of noncrystalline materials have only recently been appreciated. The application of EXAFS in catalyst studies is of particular interest, since for a number of very highly dispersed catalysts of technological importance it appears to be the only method capable of yielding detailed structural information.

EXAFS is concerned with the ejection of an inner core electron from an atom as a result of X-ray absorption. The ejected electron (photoelectron) is characterized by a wave vector K, which is given by the equation

$$K = (2mE)^{1/2}/\hbar \tag{4.1}$$

where m is the mass of the electron, $\hbar$ is Planck's constant divided by 2π, and E is the kinetic energy of the photoelectron. The energy E is the difference between the X-ray energy and a threshold energy associated with the ejection of the electron. The threshold energy depends on the particular absorption edge in question.

In the analysis of EXAFS data, it is useful to consider a function $\chi(K)$ defined by the equation

$$\chi(K) = \frac{\mu - \mu_0}{\mu_0} \tag{4.2}$$

where μ and μ_0 are atomic absorption coefficients. The coefficient μ refers to absorption by an atom in the material of interest, whereas μ_0 refers to absorption by an atom in the free state; both are functions of K. The difference between the experimental absorption coefficient at a given energy and a value at the same energy from an average background line through the data, i.e., the fluctuation, is proportional to the numerator of Eq. 4.2.

The abrupt increase in absorption coefficient at the edge is commonly called the absorption jump (or step height) and is proportional to the denominator μ_0 of Eq. 4.2. To account for the dependence of μ_0 on K, we determine a hypothetical absorption jump as a function of energy beyond the edge by employing empirical corrections based on the known energy dependence of the absorption coefficient of the element on either side of the edge. A value of $\chi(K)$ at a particular value of K is then determined by dividing the fluctuation by the hypothetical absorption jump *(18, 19)*.

Theories of EXAFS *(20–24)*, based on the scattering of ejected photoelectrons by atoms in the coordination shells surrounding the central absorbing atom, give an expression for $\chi(K)$ of the form

$$\chi(K) = \sum_j A_j(K) \sin[2KR_j + 2\delta_j(K)] \tag{4.3}$$

where the summation extends over j coordination shells. In this expression, R_j is the distance from the central absorbing atom to atoms in the jth coordination shell and $2\delta_j(K)$ is the phase shift. The factor $A_j(K)$ is an amplitude function for the jth shell, defined by the expression

$$A_j(K) = \frac{N_j}{KR_j^2} F_j(K) \exp(-2K^2\sigma_j^2) \tag{4.4}$$

where N_j is the number of atoms in the jth shell, σ_j is the root mean square deviation of the interatomic distance about R_j, and $F_j(K)$ is a factor accounting for electron backscattering and inelastic scattering *(19)*.

Fourier transformation of EXAFS data yields a function $\phi_n(R)$

$$\phi_n(R) = \left(\frac{1}{2\pi}\right)^{1/2} \int_{K_{min}}^{K_{max}} K^n \chi(K) \exp(-2iKR)\, dK \tag{4.5}$$

where R is the distance from the absorber atom. The function $\chi(K)$ is multiplied by the factor K^n before the transform is taken. The transform function has real and imaginary parts. Only the magnitude of the transform, which is everywhere positive, is considered here. It exhibits a series of peaks at $R = R_j' = R_j - a_j$. The deviation a_j from a value of R_j corresponding to a particular coordination shell arises from the phase shift in Eq. 4.3. The factor K^n in the integral is used to weight the data according to the value of K, for reasons which have been discussed elsewhere *(25)*. In practice, transforms with $n = 1$ and $n = 3$ have been used routinely. The limits K_{min} and K_{max} of the integral are the minimum and maximum values of K at which data are obtained.

When a Fourier transform is taken over a finite range of a variable, as in Eq. 4.5, a termination error is introduced into the function resulting from the transform *(26,27)*. Procedures for decreasing effects of termination errors are employed routinely in the analysis of EXAFS data *(19,28)*.

In the treatment of EXAFS data, it is useful to obtain an inverse Fourier transform of the function $\phi_n(R)$ over a limited range of R. This procedure determines the contribution to EXAFS arising from shells of atoms within that range of R. If we consider a range of R from $R'_1 - \Delta R$ to $R'_1 + \Delta R$, the inverse transform is given by the integral

$$\left(\frac{2}{\pi}\right)^{1/2} \int_{R'_1 - \Delta R}^{R'_1 + \Delta R} \phi_n(R) \exp(2iKR)\, dR \qquad (4.6)$$

If the range of R contains only one coordination shell of atoms of a single kind, the inverse transform corresponds to the product of K^n and a single term in Eq. 4.3, that is, a term for a particular value of j.

If the shell contains two types of atoms (e.g., two types of nearest neighbor atoms in the first coordination shell), the inverse transform corresponds to a sum of two terms in Eq. 4.3 multiplied by K^n. In either case, the parameters in Eqs. 4.3 and 4.4 are determined by fitting the product of K^n and the EXAFS function of Eq. 4.3 (with the appropriate number of terms) to the inverse transform function which is derived from the EXAFS data. The fitting is accomplished by means of an iterative least squares procedure *(19,28–30)*.

The analysis of EXAFS data is aided significantly by the use of parameters derived from experiments on standard materials to limit the number of unknowns in the system of interest. For example, in the analysis of EXAFS data on dispersed metal clusters containing atoms of only one metallic element, we begin by analyzing data on the pure bulk metal. If we limit our analysis to the first coordination shell of metal atoms ($j = 1$), the value of ΔR employed in obtaining the inverse transform is chosen to include only those metal atoms in the first coordination shell.

The amplitude function of Eq. 4.4 is obtained from the values of the maxima and minima in the inverse transform function. For values of K other than those corresponding to maxima or minima, the values of the amplitude function are obtained by interpolation. The phase shift is then determined by using the least squares fitting procedure. Since N_1 and R_1 of Eq. 4.4 are known for the bulk metal, it is possible to determine the product of $F_1(K)$ and

$\exp(-2K^2\sigma_1^2)$ from the amplitude function. For dispersed clusters of the same metal, the corresponding product is written as $F_1(K)\exp(-2K^2\sigma_1^2)$ $\exp(-2K^2\Delta\sigma_1^2)$, where $\Delta\sigma_1^2$ is the difference between σ_1^2 values for the dispersed metal clusters and the bulk metal.

Note that $2\delta_1(K)$ and $F_1(K)$ are assumed to be the same for the clusters and the bulk metal *(19)*. When these assumptions are made, application of the iterative least squares fitting procedure to the first shell EXAFS function derived from experimental data on the dispersed metal clusters yields values of N_1 and R_1 for the clusters. Also, one obtains the quantity $\Delta\sigma_1^2$.

In the analysis of EXAFS data on bimetallic clusters, we consider two EXAFS functions, one for each component of the clusters. If the treatment is limited to contributions of nearest neighbor backscattering atoms to EXAFS, each of the functions will consist of two terms. For a bimetallic cluster composed of elements *a* and *b*, the EXAFS associated with element *a* is given by the expression

$$\{\chi_1(K)\}_a = \{\chi_1'(K)\}_a^a + \{\chi_1'(K)\}_a^b \tag{4.7}$$

Similarly, for the EXAFS associated with element *b*

$$\{\chi_1(K)\}_b = \{\chi_1'(K)\}_b^a + \{\chi_1'(K)\}_b^b \tag{4.8}$$

In these expressions, the subscript outside the braces identifies the absorber atom, while the superscript identifies the backscattering atom. The contribution $\chi_1'(K)$ of one type of backscattering atom to the total EXAFS function is given by the equation

$$\chi_1'(K) = \frac{N_1}{KR_1^2} F_1(K)\exp(-2K^2\sigma_1^2)\sin[2KR_1 + 2\delta_1(K)] \tag{4.9}$$

where the subscript 1 signifies nearest neighbor atoms. The quantities in this equation have been defined in the discussion of Eqs. 4.3 and 4.4.

EXAFS Studies of Ruthenium–Copper Clusters (31). An X-ray absorption spectrum at 100°K showing the extended fine structure beyond the *K* absorption edge of ruthenium is given in Figure 4.6 for a catalyst containing 1.0 wt% ruthenium and 0.63 wt% copper in the form of small metal clusters

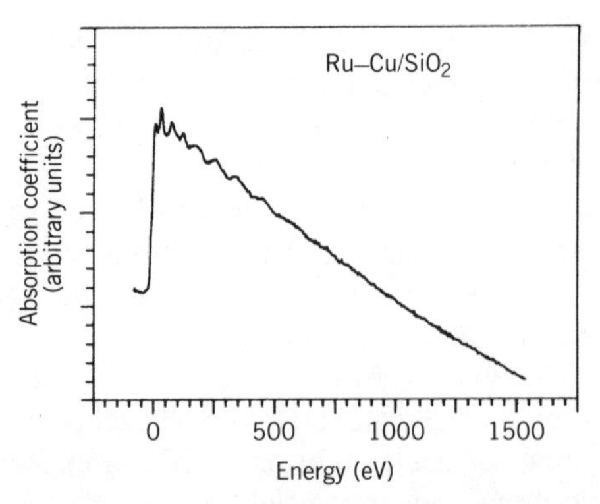

Figure 4.6 X-ray absorption spectrum of a silica-supported ruthenium–copper catalyst at 100°K in the vicinity of the *K*–absorption edge of ruthenium *(31).* (Abscissa is X-ray energy in excess of that corresponding to the edge.) (Reprinted with permission from the American Institute of Physics.)

dispersed on silica *(31).* The absorption coefficient is shown as a function of the energy of the X-ray photons in excess of the threshold energy at the absorption edge. The energy corresponds to the kinetic energy of a photoelectron ejected from the *K*-shell of a ruthenium atom.

Plots of the EXAFS function $K \cdot \chi(K)$ vs. K at 100°K for the extended fine structure beyond the ruthenium *K* edge for the ruthenium–copper catalyst and for a ruthenium on silica reference catalyst containing 1.0 wt% ruthenium are shown in the left-hand sections of Figure 4.7. The associated Fourier transforms of the functions, which were taken over the range of wave vectors 3 to 15 Å^{-1}, are shown in the middle sections of the figure *(31).* The functions $K \cdot \chi_1(K)$ shown in the right-hand sections of the figure are inverse transforms over a range of R from 1.7 to 3.1 Å. This range was chosen to include only nearest neighbor atoms of ruthenium about a ruthenium absorber atom in the case of the ruthenium catalyst, and of both ruthenium and copper about ruthenium in the case of the ruthenium–copper catalyst. The nearest neighbor atoms of ruthenium in the one case and of ruthenium and copper in the other are regarded here as constituting the first coordination shells of metal atoms about ruthenium absorber atoms. The subscript 1 in $\chi_1(K)$ signifies that the function includes only contributions to EXAFS arising from the first coordination shell of metal atoms.

The ruthenium EXAFS for the ruthenium–copper catalyst is not very different from the EXAFS for the ruthenium reference catalyst, indicating that the

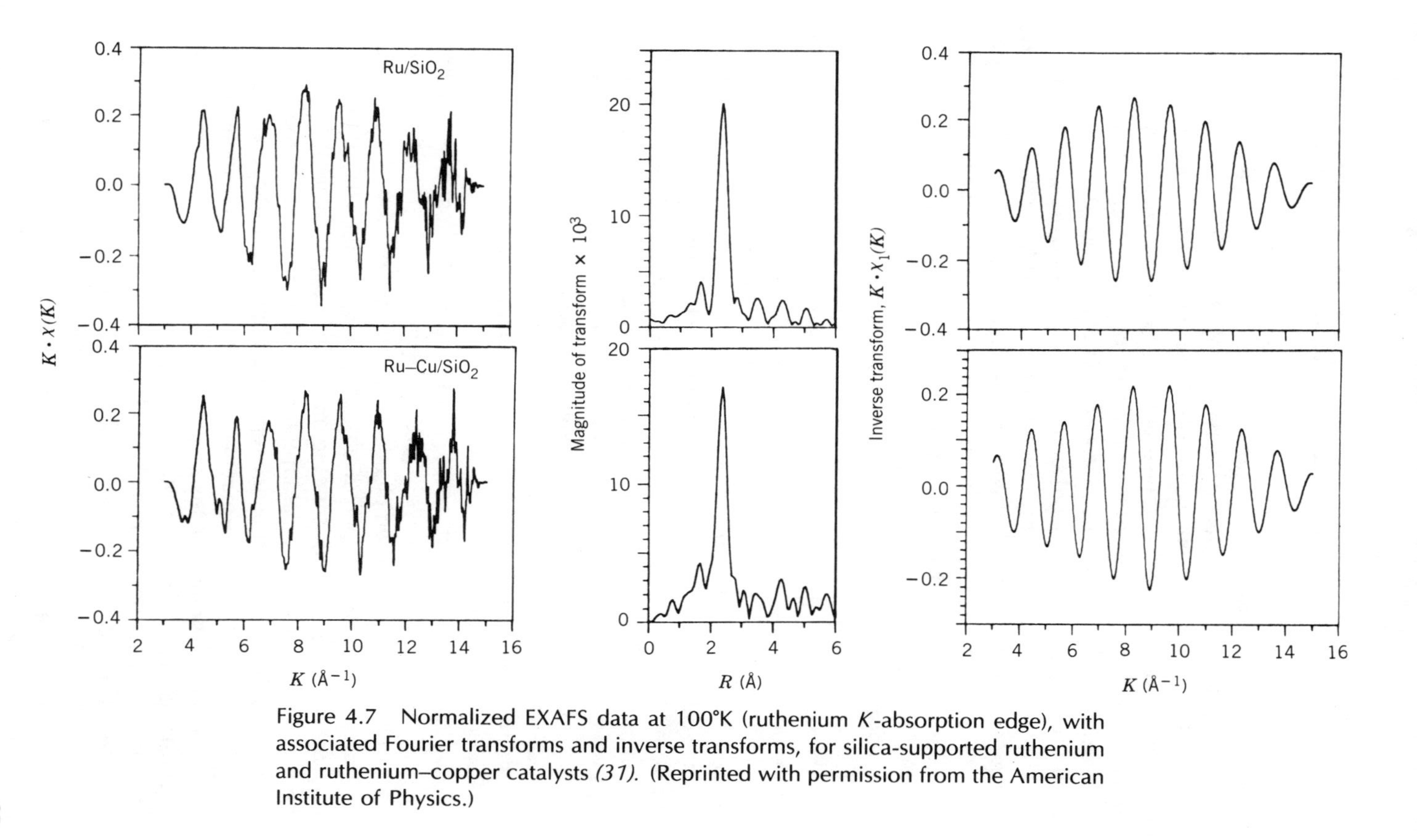

Figure 4.7 Normalized EXAFS data at 100°K (ruthenium *K*-absorption edge), with associated Fourier transforms and inverse transforms, for silica-supported ruthenium and ruthenium–copper catalysts *(31)*. (Reprinted with permission from the American Institute of Physics.)

environment about a ruthenium atom in the bimetallic catalyst is on the average not very different from that in the reference catalyst. This result is consistent with the view that a ruthenium–copper cluster consists of a central core of ruthenium atoms with the copper atoms present at the surface.

If the degree of coverage of the ruthenium by the copper is very high, as can be inferred from the chemisorption and catalysis studies already discussed, the copper atoms should be coordinated extensively to ruthenium atoms. It is emphasized that the ruthenium–copper clusters are of such a size (average diameter of 32 Å by electron microscopy) that the surface metal atoms constitute almost half of the total. Hence for a Cu/Ru atomic ratio of one, the number of copper atoms would correspond roughly to that required to form a monolayer on the ruthenium.

The copper EXAFS of the ruthenium–copper clusters might be expected to differ substantially from the copper EXAFS of a copper on silica catalyst, since the copper atoms have very different environments. This expectation is indeed borne out by experiment, as shown in Figure 4.8 by the plots of the function $K \cdot \chi(K)$ vs. K at 100°K for the extended fine structure beyond the copper K edge for the ruthenium–copper catalyst and a copper on silica reference catalyst containing 1.0 wt% copper *(31)*. The difference is also evident from the Fourier transforms and first coordination shell inverse transforms in the middle and right-hand sections of Figure 4.8. The inverse transforms were taken over the range of distances 1.7 to 3.1 Å. In the case of the copper EXAFS, we are concerned with the first coordination shell of metal atoms about a copper absorber atom. This shell consists of copper atoms alone in the copper catalyst and of both copper and ruthenium atoms in the ruthenium–copper catalyst.

A qualitative illustration of the interaction between ruthenium and copper in the ruthenium–copper catalyst is given in Figure 4.9 *(31)*. The EXAFS envelope function (or modified amplitude function) derived from the extended fine structure beyond the copper K-absorption edge for the copper reference catalyst is compared with that for the ruthenium–copper catalyst. The envelope functions were obtained from the maxima and minima in the inverse transforms in Figure 4.8 and hence represent the first coordination shells of metal atoms about copper absorber atoms. The markedly different envelope function observed for the ruthenium–copper catalyst is a consequence of the copper atoms having ruthenium atoms in addition to copper atoms as nearest neighbors. The existence of bimetallic clusters containing both ruthenium and copper atoms is clearly indicated by the data in Figure 4.9.

The effect of exposing the ruthenium–copper catalyst to oxygen at room temperature, in comparison with the effects of exposing the ruthenium and

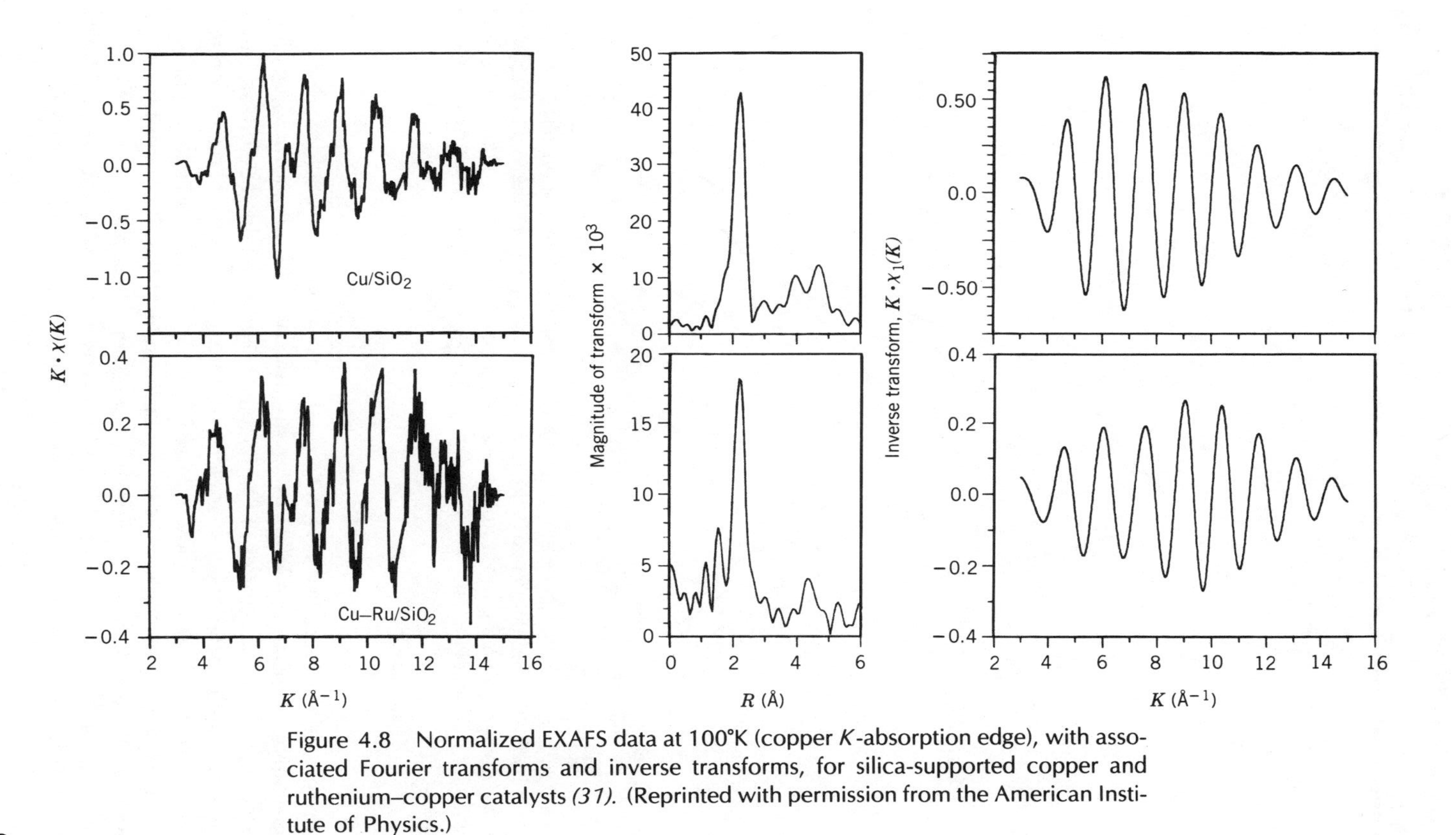

Figure 4.8 Normalized EXAFS data at 100°K (copper *K*-absorption edge), with associated Fourier transforms and inverse transforms, for silica-supported copper and ruthenium–copper catalysts *(31)*. (Reprinted with permission from the American Institute of Physics.)

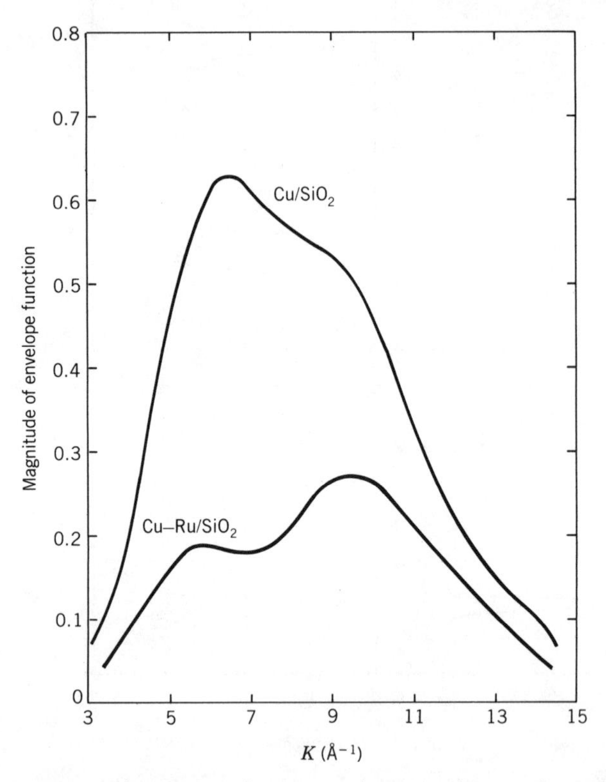

Figure 4.9 Interaction between ruthenium and copper dispersed on silica, as illustrated by the marked difference in the EXAFS envelope functions derived from EXAFS data associated with the K-absorption edge of copper in silica-supported copper and ruthenium–copper catalysts *(31)*. (Reprinted with permission from the American Institute of Physics.)

copper reference catalysts to oxygen in the same manner, provides additional qualitative information relating to the structure of the ruthenium–copper catalyst *(31)*. Fourier transforms of EXAFS data at 100°K obtained before and after exposure of the catalysts at room temperature to a stream of helium containing 1 mole% oxygen are shown, respectively, in the left- and right-hand sides of Figure 4.10. The transforms in the upper half of the figure are for EXAFS associated with the K-absorption edge of ruthenium, while those in the lower half are for EXAFS associated with the K-absorption edge of copper. Since the transforms were taken over the range of wave vectors 3 to 15 $Å^{-1}$, the peaks representative of metal–oxygen distances at values of R lower than about 2 Å are suppressed. The backscattering contribution of nearest neighbor atoms of oxygen, in contrast to that of ruthenium and copper, is high at values of K lower than 3. The suppression is not of concern to us, since we are interested in the effect of oxygen exposure on the primary peak beyond R equals 2 Å, which is the peak representing the first coordination shell of ruthenium and/or copper atoms.

In some of the transforms of EXAFS data obtained in the absence of oxygen (left-hand side of the figure), the short horizontal mark within (or slightly above) the main peak associated with the first coordination shell of metal atoms represents the magnitude of the corresponding peak from the transform (right-hand side of figure) of EXAFS data obtained in the presence of oxygen.

In the case of the ruthenium reference catalyst, the magnitude of the main peak decreases by about one-third as a result of exposure of the catalyst to oxygen. This suggests an increase in the parameter σ_1 of RuRu atomic pairs in the ruthenium clusters, and/or a structural change in the clusters leading to a decrease in the average coordination number of the ruthenium (RuRu coordination).

In contrast, the magnitude of the same peak in the transform of the ruthenium EXAFS data for the ruthenium–copper catalyst does not decrease when the data are obtained with oxygen present. This result suggests that the presence of copper tends to shield the ruthenium from the oxygen, as might be expected if the copper concentrated at the surface in the ruthenium–copper clusters *(2,10–12)*.

For the copper reference catalyst, the magnitude of the main peak in the transform is halved as a result of exposure of the catalyst to oxygen. Presumably the outer layers of the copper clusters are oxidized, leaving metallic copper cores in the interior. The oxidation of the copper apparently ceases with the formation of the outer oxide layers.

For the ruthenium–copper catalyst, the Fourier transforms of the EXAFS data associated with the copper absorption edge show a very pronounced effect of oxygen exposure, in marked contrast to the transforms associated with the ruthenium absorption edge. Thus the nature of the primary peak between 2 and 3 Å attributed to the first coordination shell of ruthenium and copper atoms is changed markedly as a result of exposure to oxygen, exhibiting two resolved maxima, with the second maximum appearing at a slightly higher value of R than is found for the original primary peak. The oxygen exposure thus has a marked preferential effect on the copper in the catalyst, which is readily understood if the copper is concentrated at the surface of

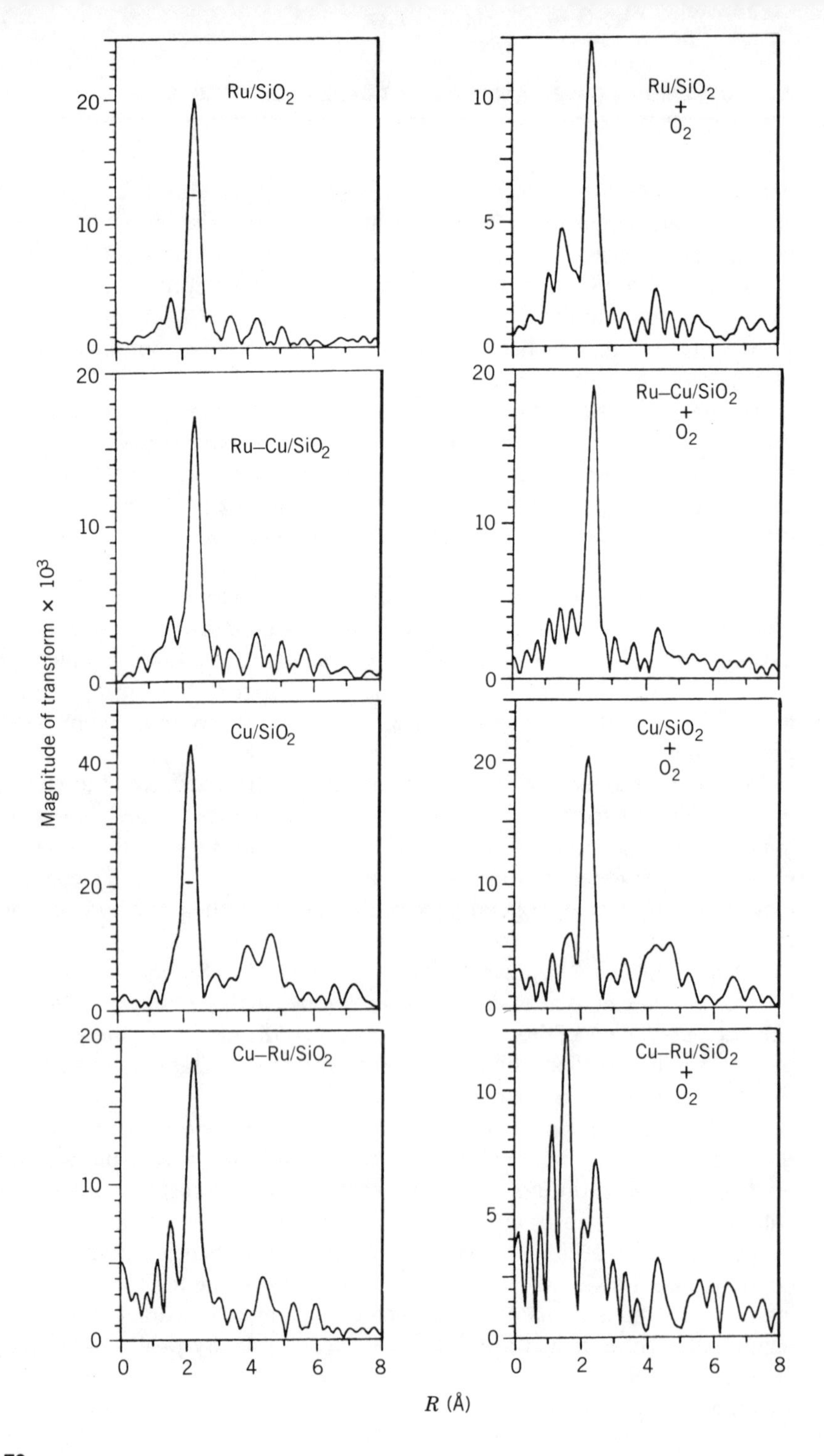

Ru/SiO2
Ru/SiO2 + O2
Ru–Cu/SiO2
Ru–Cu/SiO2 + O2
Cu/SiO2
Cu/SiO2 + O2
Cu–Ru/SiO2
Cu–Ru/SiO2 + O2
Magnitude of transform × 10^3
R (Å)
0
2
4
6
8
5
10
20
40

the ruthenium–copper clusters. Also, the fact that the Fourier transforms of the EXAFS data associated with the copper edge of the ruthenium–copper catalyst show a very different effect of oxygen exposure from that observed with the transforms for the copper reference catalyst attests to the copper being present in a very different form and environment as a result of the presence of ruthenium in the catalyst.

The discussion of EXAFS on ruthenium–copper clusters in the previous paragraphs emphasizes the qualitative aspects of the data analysis. A quantitative data analysis, yielding information on the various structural parameters of interest, has also been made and published *(31)*. Of particular interest was the finding that the average composition of the first coordination shell of ruthenium and copper atoms about a ruthenium atom was about 90% ruthenium, while that about a copper atom was about 50% ruthenium. Details of the methods involved in the quantitative analysis of EXAFS data on ruthenium–copper clusters are not considered here, since the technique is described in the following discussion of EXAFS studies on osmium–copper clusters.

EXAFS Studies of Osmium–Copper Clusters (32). Results of EXAFS studies on osmium–copper clusters dispersed on silica lead to conclusions similar to those derived for ruthenium–copper clusters. The discussion here is concerned with a catalyst with a 1:1 atomic ratio of copper to osmium. The clusters constituted 2.66 wt% of the total catalyst mass (2% Os, 0.66% Cu). The average diameter of the clusters was estimated to be about 15 Å.

In our studies of the EXAFS associated with the osmium in the clusters, we concentrated on the L_{III} absorption edge since the high energy of the K edge (73.9 KeV) was not accessible in our experiments. The nature of the X-ray absorption spectrum in the region of the L_{III} edge is illustrated in Figure 4.11 by data on pure metallic osmium at 100°K *(32)*. The abscissa of the figure is the energy of the X-ray photons. The L_{II} absorption edge is also apparent at an energy about 1.5 KeV higher than that of the L_{III} edge.

In Figure 4.12, the modified amplitude function $K \cdot (R_1^2/N_1) \cdot A_1(K)$ is

Figure 4.10 Effect of exposure of silica-supported ruthenium, ruthenium–copper, and copper catalysts to oxygen (1% oxygen in helium) at room temperature on the properties of the catalysts, as shown by Fourier transforms of EXAFS data at 100°K on the catalysts in the absence and presence of oxygen *(31)*. (Transforms in the upper half of figure are for ruthenium EXAFS; transforms in lower half are for copper EXAFS.) (Reprinted with permission from the American Institute of Physics.)

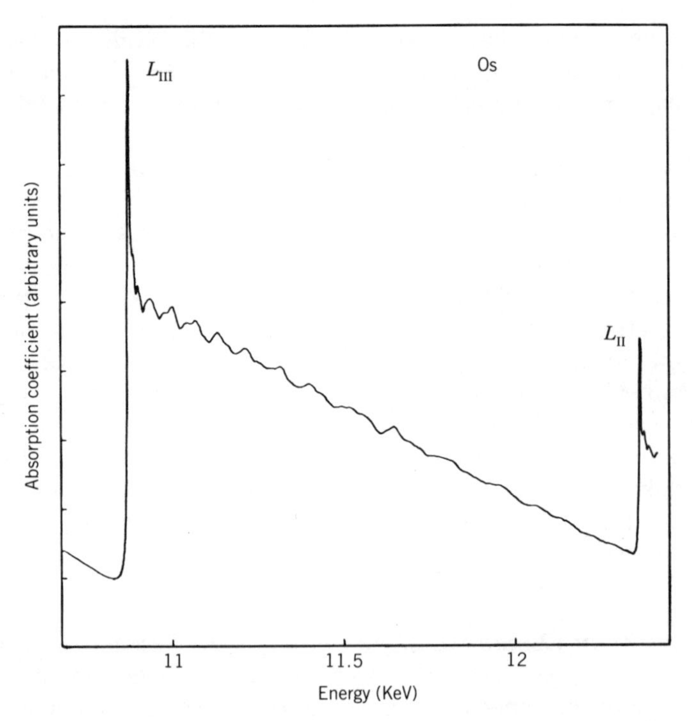

Figure 4.11 X-ray absorption spectrum of pure metallic osmium at 100°K in the region of the L_{III} and L_{II} absorption edges *(32)*. (Reprinted with permission from the American Institute of Physics.)

shown as a function of K for pure metallic osmium and copper at 100°K *(32)*. This quantity is derived from the function $K \cdot \chi_1(K)$ which, in turn, is simply the inverse Fourier transform of the function $\phi_n(R)$ over a range of R encompassing the first coordination shell of atoms. The function $\phi_n(R)$ is, of course, the Fourier transform of the initial function $K \cdot \chi(K)$ derived from the experimental EXAFS associated with the L_{III} edge of osmium or the K edge of copper.

The amplitude function $A_1(K)$ is derived from the values of the maxima and minima of the function $K \cdot \chi_1(K)$. For values of K other than those corresponding to maxima and minima, values of $A_1(K)$ are obtained by interpolation. The values of the interatomic distance R_1 for metallic osmium and copper are 2.705 and 2.556 Å, respectively. The value of 2.705 Å for metallic osmium, which has the hexagonal close-packed structure, is the average of the interatomic distance (2.735 Å) in a hexagonal layer and the distance of

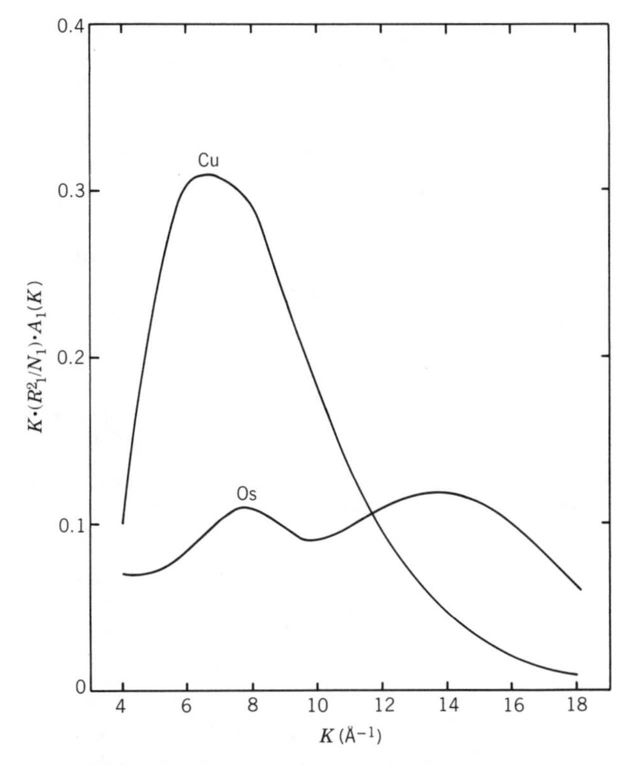

Figure 4.12 Experimentally derived values of the function $K \cdot (R_1{}^2/N_1) \cdot A_1(K)$ for pure metallic osmium and copper at 100°K *(32).* (Reprinted with permission from the American Institute of Physics.)

closest approach (2.675 Å) between two atoms in adjacent hexagonal layers *(33).* The coordination number N_1 (nearest neighbors) is 12 for both metals. For osmium, it should be noted that half of the nearest neighbors are at a distance of 2.735 Å and the other half at a distance of 2.675 Å.

From Figure 4.12 it can be seen that the modified amplitude function is very different for osmium and copper, reflecting differences in their electron scattering properties and in the parameter σ_1. As a result, osmium and copper atoms can be distinguished from each other without too much difficulty when one attempts to obtain information on the number of neighbor atoms of a given type that surround osmium or copper atoms in bimetallic clusters of these two elements.

Plots of the function $K \cdot \chi(K)$ vs. K at 100°K for the extended fine structure beyond the osmium L_{III} edge for pure metallic osmium, and for the osmium–copper clusters in the catalyst containing 2 wt% Os and 0.66 wt% Cu, are shown in the left-hand sections of Figure 4.13. The associated Fourier transforms of the functions are shown in the right-hand sections of the figure. As previously noted, the Fourier transform yields the function $\phi_n(R)$, the peaks of which are displaced from the true interatomic distances because of the phase shifts. Similar plots for the extended fine structure beyond the copper K edge for pure metallic copper and for the osmium–copper catalyst are given in our 1981 paper *(32)*.

In proceeding with the analysis of the EXAFS data, we invert Fourier transforms such as those shown in Figure 4.13 over a range of R chosen to exclude metal atoms other than nearest neighbors. This procedure yields an EXAFS function $\chi_1(K)$ in which the subscript 1 signifies that the function includes only contributions to EXAFS arising from nearest neighbor metal atoms. We begin by analyzing data on pure osmium and copper.

The amplitude function $A_1(K)$ can be obtained directly from the values of the maxima and minima of $\chi_1(K)$. Since N_1 and R_1 are known, the quantity $F_1(K) \exp(-2K^2\sigma_1^2)$ can be determined. From the values of K corresponding to the maxima and minima of $\chi_1(K)$ and also from the values at which $\chi_1(K)$ is equal to zero, the phase shift function $2\delta_1(K)$ for OsOs and CuCu atomic pairs can be determined. We have also obtained phase shifts for these pairs of atoms by adjusting theoretical values from Teo and Lee *(34)*. The adjustments were made by varying the absorption threshold energy E_0, which alters the values of the wave vector K associated with the phase shift values obtained from Teo and Lee. This procedure was continued until the expression for $\chi_1(K)$, i.e., $\chi(K)$ of Eq. 4.3 limited to the contribution of nearest neighbor backscattering atoms to EXAFS, gave a good fit to the corresponding function derived from the EXAFS data *(32)*. Changes in E_0 of -3.3 eV and -20.1 eV for OsOs and CuCu pairs, respectively, were required to obtain the best fits for metallic osmium and copper. The adjusted theoretical phase shifts for OsOs and CuCu pairs corresponding to these changes in E_0 are shown in Figure 4.14.

Phase shifts determined in this manner for OsOs and CuCu were employed to be consistent with the use of adjusted theoretical phase shifts for the osmium–copper atomic pair, which will be considered subsequently in the analysis of data on silica-supported osmium–copper clusters. For the osmium–copper pair, two phase-shift functions are necessary, depending on which of the atoms is the absorber atom and which is the backscattering atom. The two situations are distinguished by using the designation OsCu for the

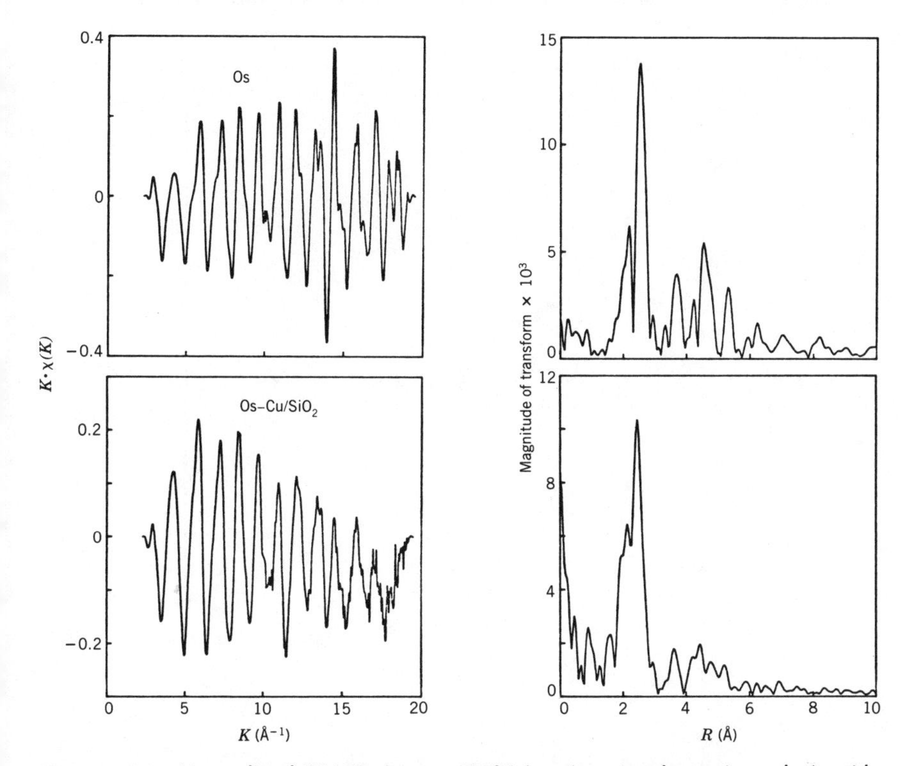

Figure 4.13 Normalized EXAFS data at 100°K (osmium L_{III} absorption edge), with associated Fourier transforms, for pure metallic osmium and a silica-supported osmium–copper catalyst containing 2 wt% Os and 0.66 wt% Cu *(32)*. (Reprinted with permission from the American Institute of Physics.)

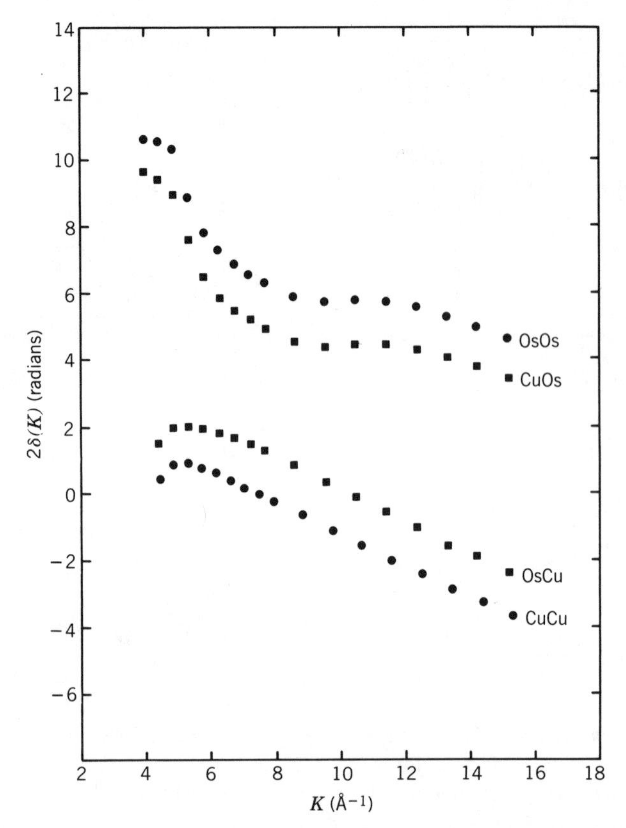

Figure 4.14 Phase shift functions for the various possible atomic pairs in the osmium–copper system *(32).* (For the osmium–copper pair, the phase shift function is designated OsCu when Os is the absorber atom and CuOs when the absorber atom is Cu.) (Reprinted with permission from the American Institute of Physics.)

osmium–copper pair when osmium is the absorber atom and CuOs when the absorber atom is copper. In the case of the osmium–copper pair we had no simple reference materials of known structure on which to obtain EXAFS data for the determination of phase shifts.

In the analysis of EXAFS data on osmium–copper catalysts, we consider two $\chi_1(K)$ functions, one for the osmium EXAFS and the other for the copper EXAFS. Each of these functions consists of two terms, one due to osmium atom neighbors and the other to copper atom neighbors about the absorber atom *(32).* Thus for the osmium EXAFS we can write the expression:

$$\left\{\chi_1(K)\right\}_{\mathrm{Os}} = \left\{\chi_1'(K)\right\}_{\mathrm{Os}}^{\mathrm{Os}} + \left\{\chi_1'(K)\right\}_{\mathrm{Os}}^{\mathrm{Cu}} \tag{4.10}$$

Similarly, for the copper EXAFS, we have

$$\left\{\chi_1(K)\right\}_{\mathrm{Cu}} = \left\{\chi_1'(K)\right\}_{\mathrm{Cu}}^{\mathrm{Cu}} + \left\{\chi_1'(K)\right\}_{\mathrm{Cu}}^{\mathrm{Os}} \tag{4.11}$$

In both expressions, the subscript outside the braces identifies the absorber atom and the superscript identifies the backscattering atom. The contribution $\chi_1'(K)$ of one type of backscattering atom to the total EXAFS function is given by Eq. 4.9.

For the osmium EXAFS, the first term in Eq. 4.10 represents the contribution of osmium backscattering atoms. In this term, the quantity N_1 represents the number of nearest neighbor osmium atoms about an osmium absorber atom and R_1 represents the distance between the osmium atoms. The phase shift function $2\delta_1(K)$ is that for an OsOs atomic pair. The quantity $F_1(K)$ $\exp(-2K^2\sigma_1^2)$ differs from the analogous quantity for pure metallic osmium by a factor $\exp(-2K^2\Delta\sigma_1^2)$, where $\Delta\sigma_1^2$ is the difference between the value of σ_1^2 for the OsOs pair in the osmium–copper catalyst and the value for the same pair in the pure metallic osmium. Note that the quantity $F_1(K)$ $\exp(-2K^2\sigma_1^2)$ for the pure metallic osmium is known from the analysis of EXAFS data on it, as indicated earlier.

The second term of Eq. 4.10 represents the contribution of copper backscattering atoms to the osmium EXAFS. In this term, N_1 is the number of nearest neighbor copper atoms about an osmium absorber atom, and R_1 is the distance between an osmium atom and a neighboring copper atom. The phase shift function $2\delta_1(K)$ in this case is the function for the OsCu atomic pair. The quantity $F_1(K)$ $\exp(-2K^2\sigma_1^2)$ now differs from the corresponding quantity for the pure metallic copper by a factor $\exp(-2K^2\Delta\sigma_1^2)$, where $\Delta\sigma_1^2$ is the difference between the value of σ_1^2 for the OsCu pair in the osmium–copper catalyst and the value for a CuCu pair in the pure metallic copper. Again, it is noted that the quantity $F_1(K)$ $\exp(-2K^2\sigma_1^2)$ for the pure metallic copper is known from the analysis of the EXAFS data on this material. The foregoing consideration of the quantities affecting the osmium EXAFS of osmium–copper catalysts applies in a similar manner to the copper EXAFS.

To continue the analysis of EXAFS data on osmium–copper catalysts, it is necessary to have phase shifts for OsCu and CuOs. The following procedure is used to obtain the phase shifts. From the paper of Teo and Lee *(34),* a phase shift function is obtained for either OsCu or CuOs *(32).* The phase shift for the other is then determined from the relation

$$2\delta^{\mathrm{OsCu}} + 2\delta^{\mathrm{CuOs}} = 2\delta^{\mathrm{OsOs}} + 2\delta^{\mathrm{CuCu}} \quad (4.12)$$

which follows simply from the additivity of the absorber and backscatterer components of phase shift functions *(34,35).* Since the phase shift functions for OsOs and CuCu are known from the EXAFS data on the pure metals, the sum of the phase shifts for OsCu and CuOs is set. It is then a question of determining how the sum is to be divided between OsCu and CuOs.

The approach adopted amounts to a trial and error procedure in which a series of values is chosen for OsCu and CuOs subject to the constraint of Eq. 4.12. For each set of trial phase shift functions, Eqs. 4.10 and 4.11 for the function $\chi_1(K)$, incorporating expressions of the form of Eq. 4.9 for the various $\chi_1'(K)$ terms, are fit to the corresponding functions derived from the osmium and copper EXAFS data on the osmium–copper catalyst. The fitting exercise yields values of various structural parameters, including the distance between an osmium atom and a copper atom (nearest neighbor atoms). For a given set of phase shift functions for OsCu and CuOs, limited only by the constraint of Eq. 4.12, this distance as derived from the osmium EXAFS will not in general be equal to the distance derived from the copper EXAFS.

We adopt the additional criterion that the distance between nearest neighbor atoms of osmium and copper must have the same value when derived from either the osmium or copper EXAFS. The phase shift functions for OsCu and CuOs which yield this result are then taken as the correct pair. The functions which are shown for OsCu and CuOs in Figure 4.14 were determined in this manner.

In arriving at these functions, we selected a number of trial phase shift functions for CuOs, including the theoretical value from Teo and Lee *(34)* and a series of altered theoretical values obtained by varying the threshold absorption energy E_0. By using Eq. 4.12, the corresponding trial phase shift functions were obtained for OsCu. Structural parameters were determined for each set of phase shift functions by means of the previously mentioned data-fitting procedure. We note that each set of trial phase shift functions may be characterized by a phase shift adjustment parameter ΔE_0, which is simply the change in threshold energy used to alter the phase shift function for CuOs from the value obtained from Teo and Lee.

Values of structural parameters, including the average composition of the first coordination shell of atoms about an osmium atom or a copper atom in the catalyst and the distance between nearest neighbor atoms of osmium and copper, are shown as a function of ΔE_0 in the upper two fields of Figure 4.15 *(32).* If we direct our attention to the second field from the top of the figure, we note the two segmented lines showing distance as a function of ΔE_0. The line labeled CuOs corresponds to distance values determined from the EXAFS associated with the copper absorption edge. The line labeled OsCu corresponds to distance values determined from the EXAFS associated with the osmium absorption edge.

The distance values, as expected, are sensitive to the phase shift functions employed, and are different for the osmium and copper EXAFS, except for the set of phase shift functions corresponding to the point of intersection of the lines. The latter are the functions shown for OsCu and CuOs in Figure 4.14 and are characterized by a CuOs phase shift adjustment parameter ΔE_0 approximately equal to -4 eV. The corresponding OsCu phase shift adjustment parameter ΔE_0 is approximately -18 eV.

It is interesting to note that the value of the E_0 adjustment for CuOs is very close to the E_0 adjustment (-3.3 eV) for OsOs required in the use of the Teo and Lee phase shift functions to fit the EXAFS results on pure metallic osmium. Similarly, the value of the E_0 adjustment for OsCu is very close to the adjustment (-20.1 eV) for CuCu required to fit the EXAFS results on pure metallic copper. Thus, for the system of interest here, it appears that the adjustments to the theoretical phase shift functions are concerned primarily with the backscattering atom.

From Figure 4.15 we obtain for the osmium–copper interatomic distance in the catalyst a value of 2.675 Å, which is about 0.05 Å larger than the sum of the metallic radii of osmium and copper *(36).* The values for the osmium–osmium and copper–copper nearest neighbor distances in the catalyst, which are not shown in the figure, are 2.680 Å and 2.550 Å, respectively. The uncertainty is estimated to be 0.01–0.02 Å in the determination of these interatomic distances *(19,32).*

The values of the osmium–osmium and copper–copper distances were insensitive to the phase shift functions employed for CuOs and OsCu over the range of ΔE_0 values for the CuOs phase shift adjustment parameter shown in Figure 4.15. The ranges of variation found for the osmium–osmium and copper–copper distances were, respectively, only 0.014 Å (2.674 to 2.688 Å) and 0.005 Å (2.549 to 2.554 Å). In both cases, but especially for the osmium–osmium pair, the distances appear to be smaller than the corresponding distances in metallic osmium and copper, which are 2.705 and 2.556 Å, respectively. As indicated earlier, the value of 2.705 Å for metallic osmium is the average of the interatomic distance (2.735 Å) in a hexagonal layer and the distance of closest approach (2.675 Å) between two atoms in adjacent

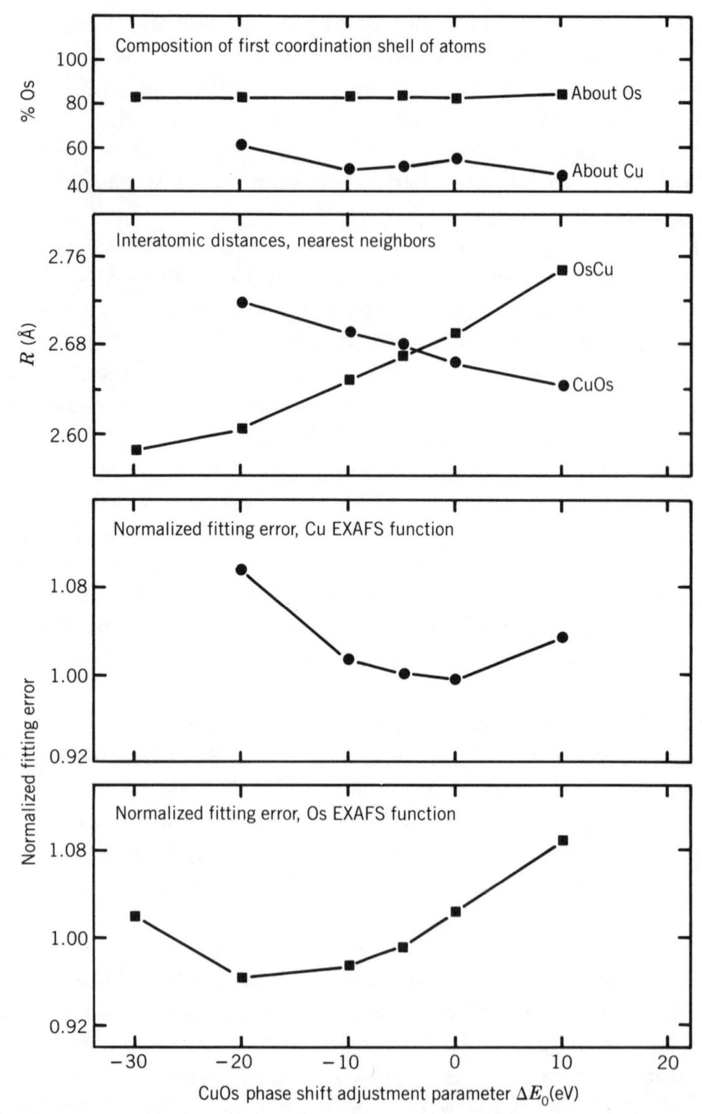

Figure 4.15 Effect of phase shift variations, as characterized by a CuOs phase shift adjustment parameter ΔE_0, on values of structural parameters obtained for a silica-supported osmium–copper catalyst containing 2 wt% Os and 0.66 wt% Cu *(32)*. (Fitting errors in lower two fields of figure provide a measure of the quality of data fit obtained.) (Reprinted with permission from the American Institute of Physics.)

hexagonal layers *(33)*. It may be noted that the osmium–osmium distance in the bimetallic clusters is very close to the smaller of the two distances for the pure osmium.

From the uppermost field of Figure 4.15 the compositions of the first coordination shells of atoms about osmium and copper atoms in the catalyst, corresponding to the osmium–copper distance of 2.675 Å determined in the manner outlined in the previous paragraphs, are 83 and 51% osmium, respectively. The composition values are not very sensitive to the value of the CuOs phase shift adjustment parameter ΔE_0 employed in the data-fitting analysis. The results indicate that the nearest neighbors about an osmium atom are predominantly osmium atoms, whereas about a copper atom they are more nearly equally distributed among osmium and copper atoms.

Information on various $\Delta\sigma_1^2$ quantities is also obtained from the data-fitting analysis. The values of σ_1^2 for the osmium–osmium and copper–copper atomic pairs in the osmium–copper catalyst are greater than the σ_1^2 values for the same atomic pairs in the corresponding pure metals by 0.0017 Å^2 and 0.0012 Å^2, respectively. The σ_1^2 value for the copper–osmium pair in the catalyst is greater than the values for the osmium–osmium and copper–copper pairs in the pure metals by 0.0033 and 0.0007 Å^2, respectively.

In the two lowest fields of Figure 4.15 we show the fitting error, or standard deviation of fit, resulting when Eqs. 4.10 and 4.11 for the function $\chi_1(K)$ are fitted to the corresponding functions derived from the osmium and copper EXAFS data on the catalyst. The fitting error, which is shown for each value of the CuOs phase shift adjustment parameter ΔE_0 characterizing a particular set of trial phase shift functions for CuOs and OsCu, has been normalized to the value at which ΔE_0 is equal to -4 eV. The normalized fitting error is seen to pass through a minimum for both the osmium and copper EXAFS, although the minimum does not occur exactly at the value of ΔE_0 (-4 eV) which corresponds to the most reasonable physical situation. We recall that the values obtained for the osmium–copper interatomic distance from the osmium and copper EXAFS data are equal only when ΔE_0 equals -4 eV.

Since one can achieve fits as good or slightly better than the fit corresponding to the most reasonable physical situation, the importance of having an independent physical criterion becomes very clear. It should be noted that the data fits for the osmium EXAFS were actually for the function $K^2\chi_1(K)$, while those for the copper EXAFS were for the function $K^3\chi_1(K)$. This unimportant procedural detail was used simply as a way of weighting the amplitude of the function $\chi_1(K)$ to roughly the same extent over the whole range of K investigated. Different weighting factors are used for the osmium and the copper EXAFS because of the differences in the shapes of the amplitude functions.

In Figures 4.16 and 4.17 the uppermost fields (labeled *a*) illustrate the quality of fit of values of the function $K^n\chi_1(K)$, represented by the points, to the corresponding function (solid line) derived from the EXAFS data *(32)*. The points were calculated for values of structural parameters corresponding to $\Delta E_0 = -4$ eV in Figure 4.15. For the osmium EXAFS in Figure 4.16 the function fitted was $K^2\chi_1(K)$, while for the copper EXAFS in Figure 4.17 it was $K^3\chi_1(K)$. The fits are excellent except at very low K values. The fits can be improved at the very low K values by modification of the details of the phase shift functions, but there is very little effect of such a modification on the values of the structural parameters obtained.

In the lower two fields of Figures 4.16 and 4.17, the points represent the separate contributions of nearest neighbor copper and osmium backscattering atoms (fields *b* and *c*, respectively) to the osmium and copper EXAFS for the osmium–copper catalyst.

In summary, the results of the EXAFS studies on osmium–copper clusters indicate that the osmium atoms in the clusters are coordinated predominantly to other osmium atoms, while the copper atoms are extensively coordinated to both copper and osmium atoms. The results are thus very similar to those obtained for ruthenium–copper clusters. Since the copper atoms appear to have essentially equal numbers of copper and osmium atoms as nearest neighbors, it seems reasonable to conclude that the osmium–copper clusters consist of small patches or multiplets of copper atoms located on the surface of the osmium.

The size of the osmium–copper clusters of interest in the catalyst considered here is such that the number of metal atoms which could be present in a full surface layer is significantly higher than the number that would be located in the interior core. For a stoichiometry of one copper atom per osmium atom, there are, then, too few copper atoms to form a complete surface layer around the osmium. It should be realized that parameters derived from the EXAFS data on the osmium–copper clusters are average values, since there is very likely a distribution of cluster sizes *(9)* and compositions in a silica-supported osmium-copper catalyst.

X-Ray Absorption Threshold Resonance Studies (32). From a study of X-ray absorption threshold resonances associated with L_{III} or L_{II} edges, one can obtain information on electronic transitions from a core level, $2p_{3/2}$ or $2p_{1/2}$, respectively, to vacant *d* states of the absorbing atom *(37,38)*. The

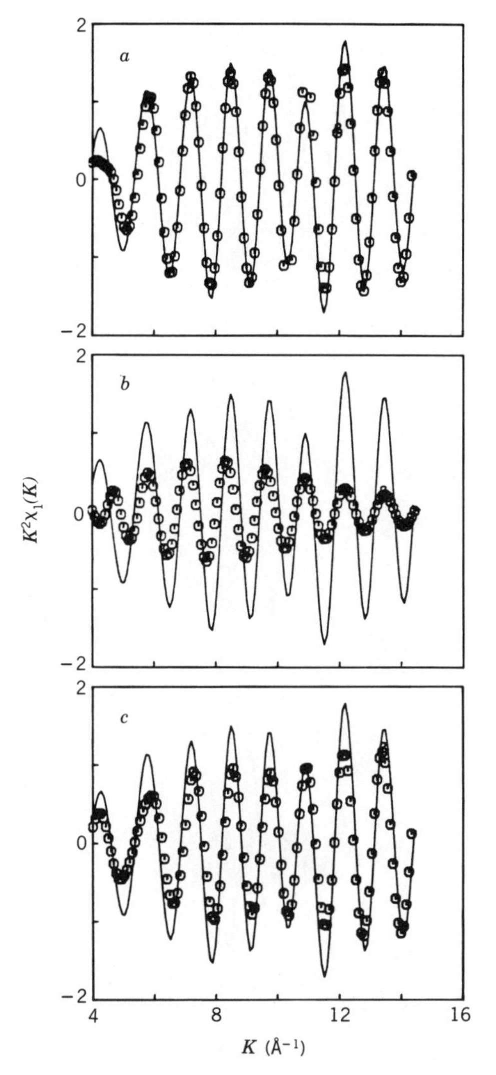

Figure 4.16 Contributions of nearest neighbor copper and osmium backscattering atoms (points in fields *b* and *c*, respectively) to the EXAFS associated with the osmium L_{III} absorption edge of a silica-supported osmium–copper catalyst containing 2 wt% Os and 0.66 wt% Cu *(32)*. (Points in field *a* show how the individual contributions combine to describe the experimental EXAFS represented by the solid line.) (Reprinted with permission from the American Institute of Physics.)

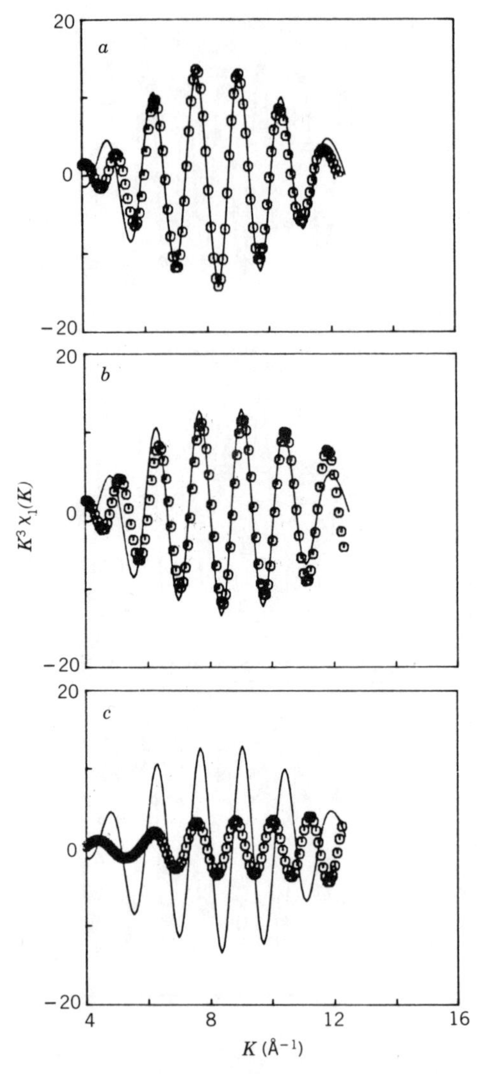

Figure 4.17 Contributions of nearest neighbor copper and osmium backscattering atoms (points in fields *b* and *c*, respectively) to the EXAFS associated with the copper *K* absorption edge of a silica-supported osmium–copper catalyst containing 2 wt% Os and 0.66 wt% Cu *(32)*. (Points in field *a* show how the individual contributions combine to describe the experimental EXAFS represented by the solid line.) (Reprinted with permission from the American Institute of Physics.)

electronic transitions are sensitive to the chemical environment of the absorbing atom *(39)*. For silica-supported osmium–copper catalysts, the magnitude of the absorption threshold resonance associated with the osmium atom is decreased by the presence of the copper. This effect is illustrated in Figure 4.18 for the L_{III} absorption edge of osmium. The absorption coefficient μ_N is a normalized absorption coefficient determined by a procedure described elsewhere *(39)*.

In the upper half of the figure, the left-hand section compares the resonance for a silica-supported osmium catalyst containing 1 wt% osmium with that for pure metallic osmium. The magnitude of the resonance is higher for the osmium dispersed on the support, the extent of increase being indicated by the difference spectrum in the lower left-hand section of the figure. This effect is similar to the results we reported for iridium and platinum dispersed on an alumina support *(39)*.

In the upper right-hand section of Figure 4.18, the magnitude of the resonance for a silica-supported osmium–copper catalyst containing 1 wt% osmium and 0.33 wt% copper (1:1 atomic ratio of copper to osmium) is compared with that for pure metallic osmium. The magnitude of the resonance for the osmium in the supported osmium–copper clusters is again higher than that for the pure metallic osmium, the extent of increase being indicated again by the difference spectrum in the lower right-hand section of the figure. However, the increase in this case is about 30% lower than is observed when the supported osmium alone is compared with pure metallic osmium; that is, the area under the difference spectral line in the lower right-hand section is about 30% smaller than the area under the corresponding spectral line in the lower left-hand section of the figure.

For the catalyst containing osmium alone on silica, the osmium clusters behave as if they are more electron deficient than pure metallic osmium; that is, there appear to be more unfilled *d* states to accommodate the electron transitions from the $2p_{3/2}$ core level of the absorbing atom. In the silica-supported osmium–copper clusters, however, the osmium atoms appear to be less electron deficient than they are in the pure osmium clusters dispersed on silica. The presence of the copper thus appears to decrease the number of unfilled *d* states associated with the osmium atoms. This observation is the first that we have made regarding the electronic interaction between the components of a bimetallic cluster catalyst. Further studies of such interactions are currently in progress on other bimetallic catalysts.

The combination of X-ray absorption threshold resonance studies with EXAFS studies provides one with the capability of obtaining information on both the structural and electronic properties of catalysts. Moreover, these

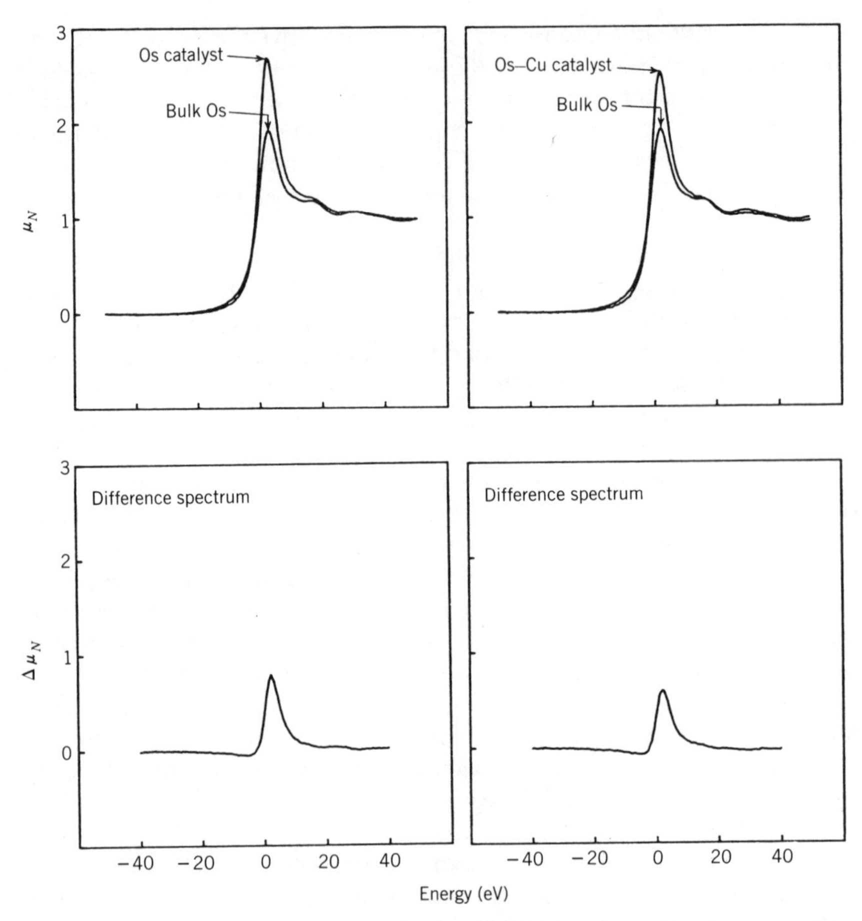

Figure 4.18 The effect of copper on the threshold X-ray absorption resonance associated with the L_{III} absorption edge of the osmium in a silica-supported catalyst *(32)*. (Reprinted with permission from the American Institute of Physics.)

types of studies have the very desirable feature that information can be obtained in an environment of the kind actually encountered in catalysis.

4.2 PLATINUM–IRIDIUM CLUSTERS

Bimetallic clusters of platinum and iridium can be prepared by coimpregnating a carrier such as silica or alumina with an aqueous solution of chloroplatinic and chloroiridic acids *(3,4)*. After the impregnated carrier is dried and possibly calcined at mild conditions (250–270°C), subsequent treatment in flowing hydrogen at elevated temperatures (300–500°C) leads to formation of the bimetallic clusters.

4.2.1 Characterization by Chemical Probes

Data on the chemisorption of hydrogen at room temperature on platinum–iridium clusters dispersed on alumina and silica are shown in Figures 4.19 and 4.20 as a function of the amount of platinum and iridium in the catalyst *(4)*. The data are for catalysts containing equal fractions by weight of platinum

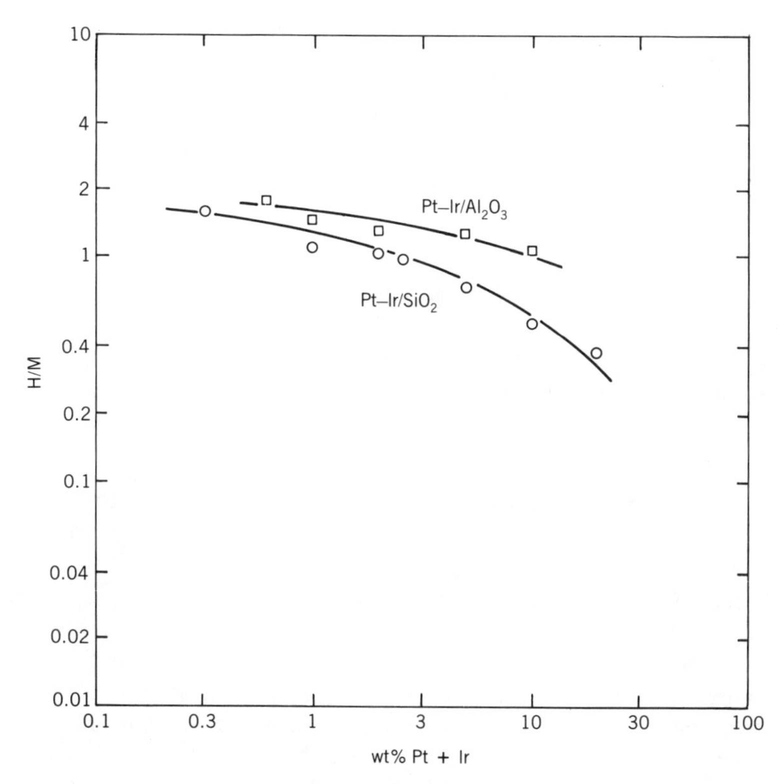

Figure 4.19 Hydrogen chemisorption, at room temperature and 10 cm Hg pressure, as a function of the total content of platinum and iridium in catalysts comprising equal weights of platinum and iridium dispersed on alumina or silica *(4)*. (Reprinted with permission from Academic Press, Inc.)

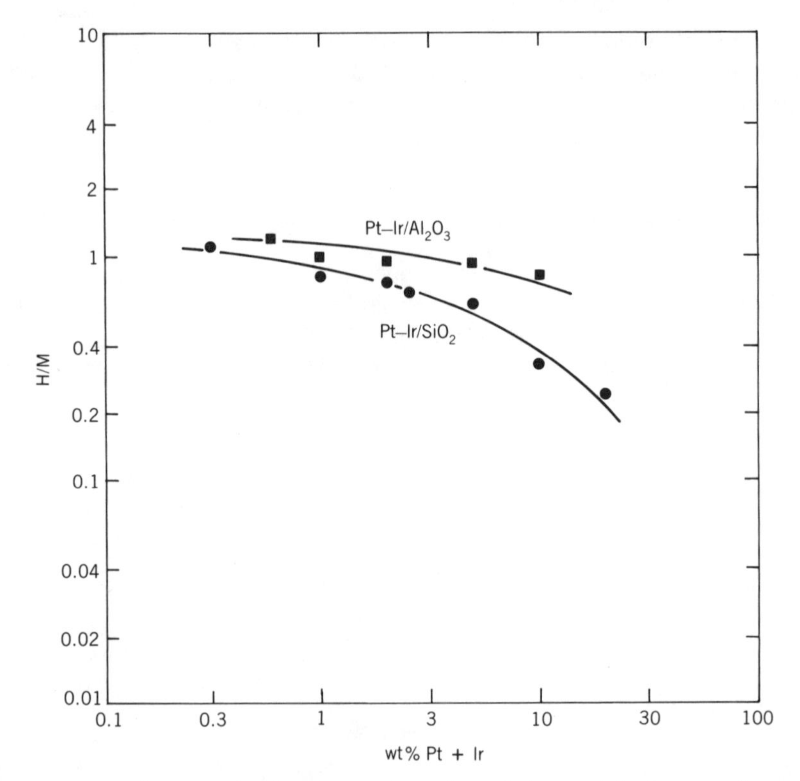

Figure 4.20 The strongly chemisorbed fraction of the total hydrogen adsorption shown in Figure 4.19 *(4)*. (Reprinted with permission from Academic Press, Inc.)

and iridium (corresponding to a Pt/Ir atomic ratio very close to unity). The surface areas of the alumina and silica used in the catalysts are approximately 200 and 300 m^2/g, respectively.

The amount of hydrogen adsorbed is expressed in terms of the quantity H/M, which represents the ratio of the number H of adsorbed hydrogen atoms to the number M of metal atoms (platinum and iridium) in the catalyst. In Figure 4.19 the data are for total hydrogen chemisorption at a pressure of 10 cm Hg. The data in Figure 4.20 represent strongly chemisorbed hydrogen, that is, the amount retained by the catalyst when the adsorption cell is evacuated at room temperature for 10 minutes (to a pressure of approximately 10^{-6} torr) after completion of the adsorption isotherm. As noted previously, in Chapters 2 and 3, this quantity is the difference between the isotherm for total hydrogen chemisorption and a second isotherm for weakly adsorbed hydrogen determined after the evacuation step.

For both total and strong chemisorption, H/M increases as the metal content of the catalyst decreases and is consistently higher for catalysts in which the platinum–iridium clusters are dispersed on alumina. As shown in Figure 4.19, the H/M values for total chemisorption frequently exceed unity. Values of H/M approaching 2 are observed at the lowest metal contents when the platinum–iridium clusters are dispersed on alumina. For strongly chemisorbed hydrogen, H/M appears to approach a limiting value near unity as the metal content is decreased to about 1 wt% or lower.

Electron microscopy data on such catalysts show the average diameters of the metal clusters to be of the order of 10 Å or lower. Clusters of this size necessarily consist almost exclusively of surface atoms. The stoichiometry of the strongly chemisorbed fraction thus appears to be close to one hydrogen atom per surface metal atom for platinum–iridium catalysts containing equal amounts of the two metals. If this stoichiometry were precisely correct, a value of H/M from Figure 4.20 would be a direct measure of the metal dispersion, that is, the ratio of surface atoms to total atoms in the metal clusters.

While a value of H/M from Figure 4.20 may be taken as an approximate measure of the metal dispersion, inspection of the data on the alumina-supported platinum–iridium catalyst with the lowest metal content indicates that the value of H/M may be as high as 1.3. Consequently, there is slightly more than one hydrogen atom per surface metal atom in the chemisorbed layer remaining after the adsorption cell is evacuated at room temperature.

If the alternative procedure of extrapolating the nearly pressure-independent region of the original adsorption isotherm back to zero pressure is employed, as discussed in Chapter 2, it is observed that the value of H/M for the alumina-supported platinum–iridium clusters in the catalyst containing 0.6 wt% metal is about 1.7. The strongly chemisorbed hydrogen determined by the method involving room temperature evacuation is approximately 75% of this value.

In Table 4.1 chemisorption data on alumina-supported platinum–iridium catalysts and related catalysts containing platinum or iridium alone show the effect of varying the temperature of calcination of the catalyst (in air or oxygen–helium mixture) on the metal dispersion *(40,41)*. Data are presented for chemisorption of carbon monoxide, hydrogen, and oxygen. The final three catalysts in the table contained more metal than the first three. They also contained 0.1 wt% Fe (enriched with ^{57}Fe) incorporated as a probe for Mössbauer spectroscopy experiments *(41)*. The presence of the iron is ignored in the discussion of the chemisorption results.

Table 4.1 Effect of Calcination Temperature on Metal Dispersion in Iridium, Platinum, and Platinum–Iridium Catalysts *(40,41)*

Catalyst[a]	Calcination Temperature, °C[b]	CO/M[c]	H/M[c]	O/M[c]
0.3% Ir[d]	270	1.8	2.5	
	500	0.3	0.5	
0.3% Ir, 0.3% Pt[d]	270	1.2	1.7	
	500	0.8	1.2	
0.3% Pt[d]	500	0.8	1.0	
1.75% Ir[e]	260		1.29	0.90
	600		0.04	0.09
1.75% Ir, 1.75% Pt[e]	260		1.00	0.90
	600		0.57	0.56
1.75% Pt[e]	260		0.86	0.78
	600		0.95	0.86

[a]Alumina was the carrier in all of the catalysts.

[b]Calcination in air or oxygen–helium mixture containing 20% oxygen. After calcination, the catalysts were reduced in hydrogen at 500°C and evacuated at 450°C prior to chemisorption experiments at room temperature.

[c]The quantities CO/M, H/M, and O/M represent, respectively, the number of carbon monoxide molecules, hydrogen atoms, and oxygen atoms chemisorbed at room temperature per atom of metal M dispersed on the alumina. The values of CO/M represent the amounts of adsorbed CO retained after the adsorption cell is evacuated at room temperature for 10 min to a pressure of approximately 10^{-6} torr subsequent to completion of the adsorption isotherm. The values of H/M and O/M were determined by extrapolation of the nearly pressure-independent regions of the adsorption isotherms to zero pressure.

[d]Data taken from Ref. 40, with permission from Elsevier Scientific Publishing Company.

[e]Data taken from Ref. 41, with permission from Academic Press, Inc. These catalysts contained 0.1 wt% Fe (enriched with ^{57}Fe) incorporated as a probe for Mössbauer spectroscopy experiments.

The values of CO/M in Table 4.1 represent the amounts of adsorbed CO determined by the method involving evacuation of the adsorption cell at room temperature following completion of the original adsorption isotherm. The H/M and O/M values were determined by the method involving back extrapolation of the nearly pressure-independent region of the original isotherm to zero pressure.

For platinum dispersed on alumina, it is generally accepted that the stoichiometry of hydrogen chemisorption corresponds to the adsorption of one atom of hydrogen per surface platinum atom *(42)*. In the chemisorption of carbon monoxide, there is evidence of some contribution of an adsorbed species requiring two platinum atoms per carbon monoxide molecule. However, the form requiring only one surface platinum atom appears to dominate *(42)*. On the basis of these considerations, one would expect the quantities CO/M and H/M to approach a maximum value of unity when the dispersion of the platinum is very high, that is, when the platinum atoms are virtually all surface atoms. A limiting value of unity for O/M might also be expected. The data in Table 4.1 are consistent with these expectations.

Increasing the temperature of calcination of the catalyst containing 1.75% platinum from 260 to 600°C had little effect on the chemisorption results, increasing the values of H/M and O/M by about 10%.

For iridium dispersed on alumina, the data on the catalysts calcined at 260–270°C indicate that more than one atom of hydrogen or one molecule of carbon monoxide is adsorbed per surface iridium atom *(40,43)*. However, the data on the quantity O/M show no evidence of values in excess of unity. In contrast with platinum catalysts, increasing the temperature of calcination of iridium catalysts to 500–600°C decreases chemisorption capacities markedly. Calcination in air or oxygen at 500–600°C leads to formation of large IrO_2 crystallites that yield large iridium crystallites on reduction *(3,4,41)*. The fraction of the iridium atoms present as surface atoms thus decreases markedly, and this, in turn, is reflected in marked decreases in the chemisorption capacity for all three gases.

For the platinum–iridium catalyst calcined at 260°C, the value of O/M differs very little from the O/M values for the corresponding platinum and iridium catalysts. The differences in the H/M values are larger, but the value for the platinum–iridium catalyst is very close to the average of the H/M values for the platinum and iridium catalysts.

For the platinum-iridium catalyst calcined at 500°C, however, the values of CO/M and H/M are significantly higher than the averages for the platinum and the iridium catalysts calcined at the same temperature. While calcination of the platinum–iridium catalyst at 500°C also leads to formation of large IrO_2 crystallites, and subsequently to large iridium or iridium-rich crystallites on reduction *(3,4)*, the deleterious effect of the calcination is clearly not so pronounced as it is in the iridium catalyst, which contains no platinum. The presence of the platinum inhibits the oxidative agglomeration of the iridium at 500°C. This effect, although less pronounced, is still observed when the calcination temperature is increased to 600°C, as indicated by the H/M and O/M values on the catalysts calcined at this temperature. This observation agrees with the view that the platinum and the iridium in the catalyst are present as bimetallic clusters rather than separate clusters of pure platinum and pure iridium *(3,4)*.

The observation that the iridium is less susceptible to oxidative agglomeration in the presence of the platinum has the important practical feature that the platinum–iridium catalyst is more stable than a pure iridium catalyst in the oxygen-containing gases employed for the regeneration of the catalysts. In general, the regeneration procedures employed with platinum–iridium catalysts will differ from those used with platinum catalysts because of the greater susceptibility of iridium to oxidative agglomeration.

The use of the ethane hydrogenolysis reaction as a chemical probe in the characterization of platinum–iridium on alumina catalysts is illustrated by the data in Figure 4.21. Rates of hydrogenolysis per iridium atom are shown over a range of temperatures for a series of catalysts containing 0.3 wt% iridium

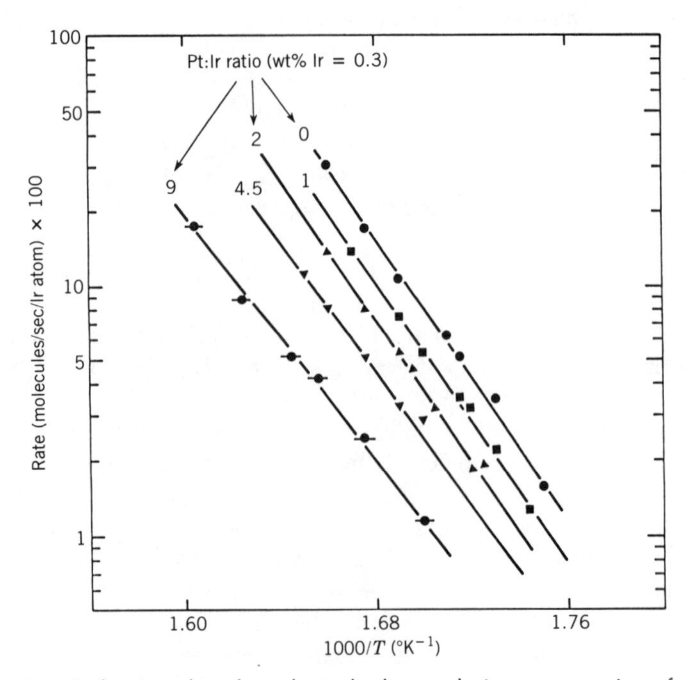

Figure 4.21 Arrhenius plots for ethane hydrogenolysis over a series of alumina-supported platinum–iridium catalysts containing 0.3 wt% iridium and variable amounts of platinum corresponding to Pt:Ir atomic ratios ranging from 0 to 9.

and variable amounts of platinum corresponding to Pt:Ir atomic ratios ranging from 0 to 9. The ratios are designated by the numbers associated with the various Arrhenius lines in the figure. The hydrogenolysis rates were determined at ethane and hydrogen partial pressures of 0.030 and 0.97 atm, respectively. As the atomic ratio of platinum to iridium increases, the hydrogenolysis rate decreases.

The effect of platinum on the hydrogenolysis activity of iridium is dramatically lower than the effect of copper on the hydrogenolysis activities of metals such as nickel, ruthenium, and osmium. Nevertheless, the effect is significant and is an indication of interaction between the platinum and iridium. The contents of platinum and iridium were maintained at a low enough level in all of the catalysts to ensure that the metal dispersion was close to unity throughout. The metal dispersion is estimated to be in the range of 0.8 to 1.0 for all of the catalysts. In this way we attempt to detect interaction between platinum and iridium in catalysts in which virtually all of the atoms of both metals are surface atoms. The hydrogenolysis activity per iridium atom in the catalyst may then be identified with the activity per surface iridium atom.

If there were no interaction between the platinum and iridium in the catalysts, that is, if the catalysts consisted simply of mixtures of platinum clusters and iridium clusters, it would be reasonable to expect the hydrogenolysis activity per iridium atom to remain constant as platinum is incorporated in the catalysts. This expectation would be based on the fact that the activity of platinum for hydrogenolysis is negligible compared to that of iridium, about five orders of magnitude lower for ethane hydrogenolysis *(44),* so that the added platinum would presumably behave as an inert diluent.

However, the data of Figure 4.21 show that the incorporation of platinum decreases the activity per iridium atom in the catalyst. A catalyst with a Pt:Ir atomic ratio of 2 (containing 0.6 wt% platinum and 0.3 wt% iridium) is about half as active for hydrogenolysis as a catalyst containing 0.3 wt% iridium alone. Increasing the Pt:Ir atomic ratio to 9 (2.7 wt% platinum and 0.3 wt% iridium) yields a catalyst with a hydrogenolysis activity about sevenfold lower than that of the iridium catalyst. These data provide a clear indication of interaction between the platinum and iridium.

There is an analogy between the platinum–iridium system and systems such as ruthenium–copper or osmium–copper in the sense that one of the components (platinum or copper) possesses a hydrogenolysis activity that is negligible relative to that of the other (iridium, ruthenium, or osmium). However, the platinum–iridium system differs in the magnitude of the effect of the inactive component on the hydrogenolysis activity of the system. The nature of the interaction between platinum and iridium could reasonably be expected to differ from the interaction between copper and ruthenium or between copper and osmium. These differences could be reflected in structural differences in the bimetallic clusters and/or differences in the influence of an electronic (ligand) effect on their hydrogenolysis activities.

4.2.2 Characterization by Physical Probes

Various types of physical probes have been employed in our studies of platinum–iridium clusters. The probes include X-ray diffraction, extended X-ray absorption fine structure (EXAFS), and Mössbauer effect spectroscopy. All of the probes have provided useful information for the characterization of the bimetallic clusters.

X-Ray Diffraction Studies. The dependence of the lattice parameter of bulk platinum–iridium alloys on composition is shown in Figure 4.22 *(4,45).* Lattice parameters are commonly obtained from X-ray diffraction measurements. For platinum–iridium catalysts, X-ray diffraction measurements provide a way of demonstrating the presence of bimetallic clusters of platinum and iridium, if the metal dispersion is not too high.

X-ray diffraction data for a silica-supported platinum–iridium catalyst containing 10 wt% platinum and 10 wt% iridium *(4)* are shown in the middle field of Figure 4.23. The catalyst was diluted with an equal weight of silica. The metal dispersion of the catalyst as determined by hydrogen chemisorption was 0.24. A portion of the diffraction pattern is shown that includes the *(220)* reflection. The upper field of the figure, which shows the corresponding part of the diffraction pattern of a physical mixture of bulk platinum and bulk iridium diluted with silica, serves as a reference. The lower field of the figure shows the same region of the diffraction pattern for a physical mixture of equal weights of platinum on silica and iridium on silica catalysts, the former containing 10 wt% platinum and the latter 10 wt% iridium. The absolute amounts of platinum and iridium in this mixture were equal to the amounts in the silica-diluted platinum–iridium catalyst sample used in obtaining the data in the middle field of the figure. The background scattering arising from the silica carrier was subtracted from the total scattering in the diffraction patterns for the catalysts.

In the diffraction patterns in Figure 4.23, the platinum–iridium bimetallic clusters exhibit a single diffraction line (middle field of figure) about midway between the lines (upper field of figure) for bulk platinum and bulk iridium. The line for the clusters is broader than the lines for the bulk metals because

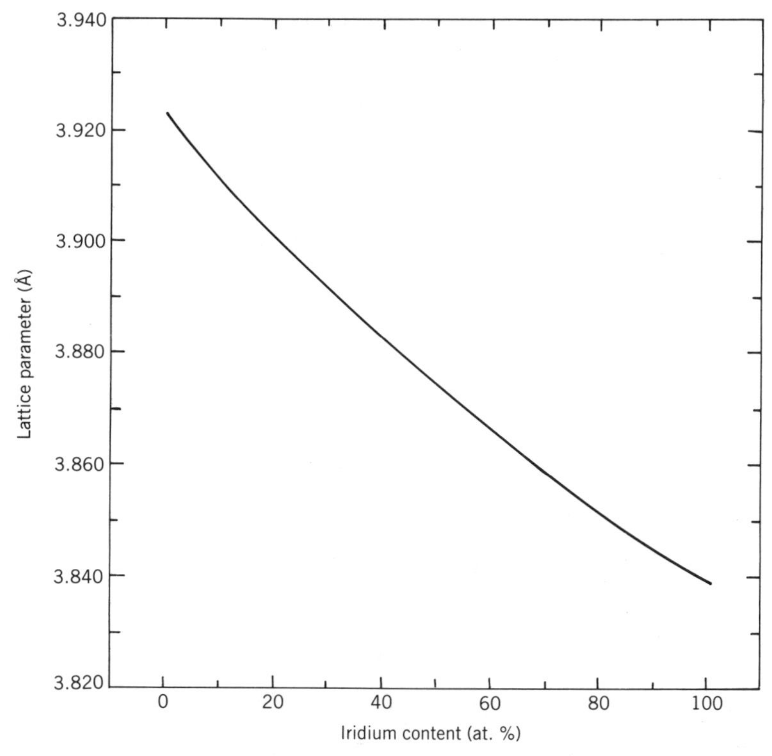

Figure 4.22 Lattice parameter of platinum–iridium solid solutions as a function of composition *(4,45)*. (Reprinted with permission from Academic Press, Inc.)

of the small size of the clusters. An average platinum–iridium cluster size of 49 Å is calculated from the line width by means of the Scherrer equation *(46)*. The lattice parameter of the platinum–iridium clusters is 3.875 Å, as determined from the positions of a number of lines in the diffraction pattern. This value corresponds to a composition of 50% iridium in Figure 4.22, in agreement with the overall composition.

The diffraction pattern of the physical mixture of silica-supported platinum clusters and iridium clusters in the lower field of Figure 4.23 consists of overlapping lines for the two individual types of clusters, and is clearly different from the pattern for the platinum–iridium bimetallic clusters. The overlapping is due to the broadened nature of the individual lines resulting from the small sizes of the individual clusters of platinum and iridium.

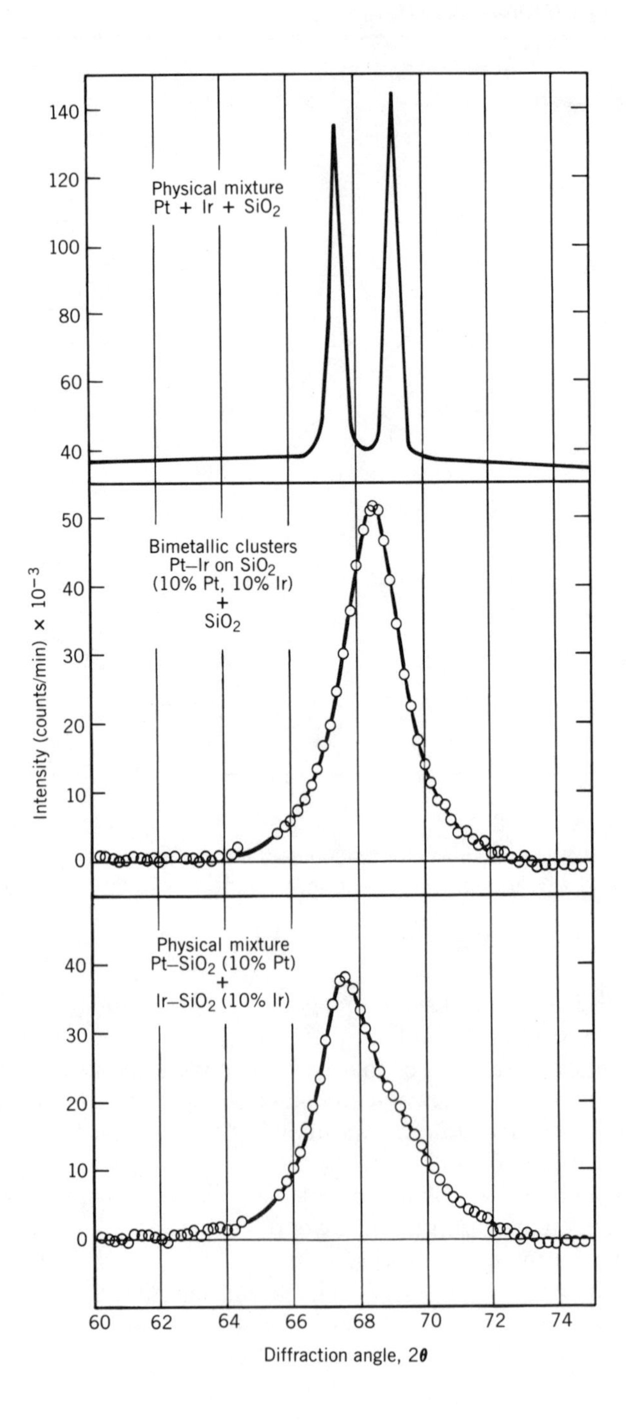
140
120
100
80
60
40
50
40
30
20
10
0
40
30
20
10
0
Physical mixture
Pt + Ir + SiO2
Bimetallic clusters
Pt–Ir on SiO2
(10% Pt, 10% Ir)
+
SiO2
Physical mixture
Pt–SiO2 (10% Pt)
+
Ir–SiO2 (10% Ir)
Intensity (counts/min) × 10⁻³
60
62
64
66
68
70
72
74
Diffraction angle, 2θ

In studies of the platinum–iridium system in the bulk, Raub and Platte *(47)* have reported a miscibility gap at temperatures lower than about 975°C. At 500°C the gap extends over the composition range from 7 to 99% iridium. Nevertheless, the results presented in Figure 4.23 indicate that it is possible to prepare platinum–iridium catalysts for which X-ray diffraction patterns do not reveal separate lines for platinum-rich and iridium-rich phases, despite the fact that the catalysts have been heated to only 500°C in their preparation. Instead, single diffraction lines are observed.

It is possible to obtain a single symmetrical line for a two-phase system over a certain range of compositions of the phases. A single, symmetrical line can be obtained by adding two Lorentzian-shaped lines centered at diffraction angles differing by as much as 0.7° and corresponding to compositions as far apart as 70% platinum and 30% platinum *(48).* However, an attempt to obtain a single, symmetrical line in this manner for a pair of phases with compositions as far apart as those indicated by the work of Raub and Platte was unsuccessful *(48).* Thus if the platinum–iridium clusters consist of two separate phases, it is concluded that the individual phases contain substantial amounts of the minor components. It is possible that the phase equilibria may be influenced by the small size of the metal clusters in the catalyst. Approximately one-fourth of the metal atoms in the platinum–iridium bimetallic clusters considered in Figure 4.23 are surface atoms.

X-ray diffraction is useful for investigating silica-supported platinum–iridium clusters with metal dispersions as high as 0.60 *(3,4).* For platinum–iridium clusters of higher dispersion, however, the application of X-ray diffraction becomes very difficult, as it does for supported metal catalysts of any kind having such high dispersions. The lines in the diffraction patterns are then extremely broad and weak, becoming indistinguishable from the background. Furthermore, for alumina-supported platinum–iridium catalysts, there is the additional complication that the diffraction pattern of the alumina interferes with that of the platinum–iridium clusters.

In the preparation of highly dispersed platinum–iridium bimetallic clusters on a silica or alumina carrier, it is important to avoid heating (calcining) the material in air at too high a temperature *(3,4).* At 500°C the iridium undergoes oxidation and agglomeration to form large crystallites of IrO_2 (greater than

---

Figure 4.23 X-ray diffraction patterns for a platinum–iridium bimetallic cluster catalyst and for reference materials consisting of physical mixtures of platinum and iridium in the form of large crystals or dispersed monometallic clusters *(4).* (Reprinted with permission from Academic Press, Inc.)

200 Å in size). On subsequent treatment in hydrogen at 500°C, the IrO_2 crystallites are reduced to metallic iridium crystallites. The material then consists of a mixture of highly dispersed platinum or platinum-rich clusters and large crystallites of iridium.

The diffraction profile for the (220) region for a silica-supported platinum–iridium catalyst calcined (heated) in air at 500°C prior to reduction in hydrogen at 500°C is given in the upper field of Figure 4.24. The catalyst con-

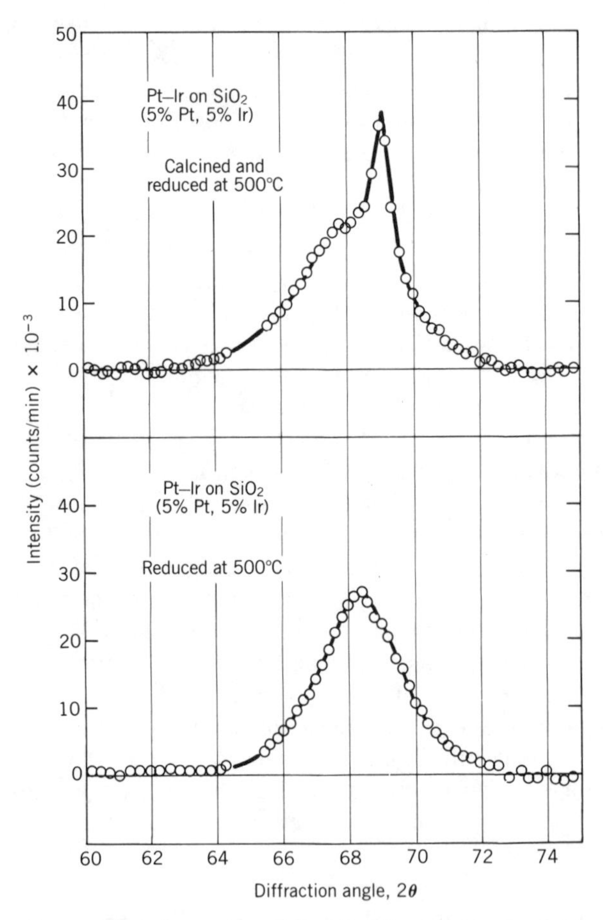

Figure 4.24 X-ray diffraction pattern showing the effect of calcining (heating) a silica-supported platinum–iridium sample in air at 500°C prior to reduction in hydrogen at 500°C, illustrating the importance of the preparative conditions in the formation of highly dispersed platinum–iridium bimetallic clusters *(4)*. (Reprinted with permission from Academic Press, Inc.)

tained 5 wt% each of platinum and iridium. There is a marked contrast between the diffraction pattern for this material and the pattern in the lower field of the figure for a sample of the material that had not been heated in air at 500°C before reduction. The pattern in the upper field consists of two overlapping lines, a broad one for dispersed platinum and a much narrower one for large iridium crystallites. The pattern in the lower field consists of a single line characteristic of bimetallic clusters of platinum and iridium. These results demonstrate the importance of preparative conditions in the formation of bimetallic clusters in the platinum–iridium system. In general, exposure to air at temperatures below about 375°C does not appear to be harmful, but contact with air at temperatures above about 450°C should be avoided *(3,4)*.

EXAFS Studies (48). An X-ray absorption spectrum at 100°K in the region of the L absorption edges of iridium and platinum is given in Figure 4.25 for a silica-supported platinum–iridium catalyst containing 10 wt% each of the individual metals *(48)*. The catalyst is the same one for which X-ray diffraction data were shown in the middle field of Figure 4.23. The X-ray absorption data in Figure 4.25 were obtained over a range of energies of the X-ray photons wide enough to include all of the L absorption edges of iridium and platinum. In the following discussion we are concerned with the extended fine structure associated with L_{III} absorption edges.

The L_{III} absorption edges of iridium and platinum are 348.5 eV apart in energy. Since the extended fine structure associated with the L_{III} edge of iridium is observable to energies of 1200–1300 eV beyond the edge, there is overlap of the EXAFS associated with the L_{III} edges of iridium and platinum for a catalyst containing both of these elements. Separating the iridium EXAFS from the platinum EXAFS in the region of overlap is a key feature in the analysis of the data, as will be seen subsequently.

The dependence of the function $K^3\chi^+(K)$ on the wave vector K (at 100°K) for the extended fine structure beyond the platinum L_{III} edge, where the function $\chi^+(K)$ is equal to the sum of the EXAFS function $\chi(K)$ and an additional function $I(K)$ arising from the overlapping iridium EXAFS *(48)*, is shown in the left-hand sections of Figure 4.26 for three platinum–iridium catalysts: Pt–Ir/SiO_2 (10 wt% Pt, 10 wt% Ir), Pt–Ir/SiO_2 (1 wt% Pt, 1 wt% Ir), and Pt–Ir/Al_2O_3 (1 wt% Pt, 1 wt% Ir). Fourier transforms of the functions taken over the range of wave vector 2–18 $Å^{-1}$ are shown in the middle sections of Figure 4.26.

The functions $K^3\chi_1^+(K)$ shown in the right-hand sections of Figure 4.26

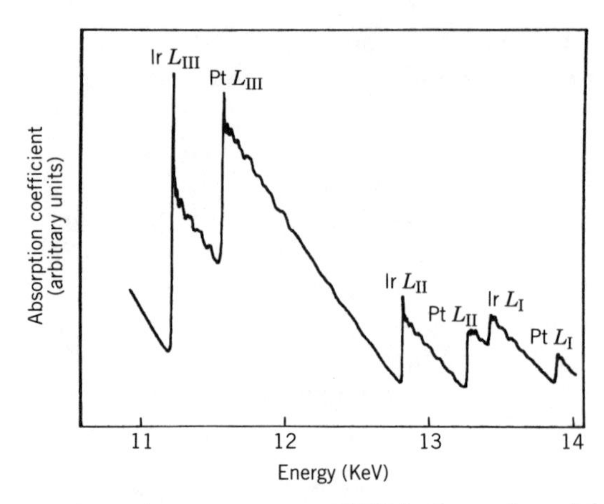

Figure 4.25 X-ray absorption spectrum at 100°K in the region of the L absorption edges of iridium and platinum for a platinum–iridium on silica catalyst containing 10 wt% each of platinum and iridium *(48)*. (Reprinted with permission from the American Institute of Physics.)

were obtained from the Fourier transforms in the middle sections by taking inverse transforms of the latter over a restricted range of R. The range 1.4–3.3 Å was chosen to limit the platinum EXAFS part of the inverse transform to contributions from nearest neighbor platinum and iridium atoms, as is signified by the use of the subscript 1 in $\chi_1(K)$ and $\chi_1{}^+(K)$. The latter function is given by the expression

$$\chi_1{}^+(K) = \chi_1(K) + I_1(K) \tag{4.13}$$

where the term I_1(K) arises from the function $I(K)$.

In the analysis of the EXAFS data, we restrict our attention to the fine structure associated with the platinum L_{III} edge, since the fine structure associated with the iridium L_{III} edge has the complication of overlapping with the platinum L_{III} edge in addition to overlapping with the EXAFS associated with the edge *(48)*. We also restrict our attention to the EXAFS resulting from nearest neighbor atoms and approach the problem by evaluating the function $I_1(K)$ introduced in Eq. 4.13. Such an evaluation requires a knowledge of the iridium L_{III} EXAFS for the platinum–iridium catalyst in the region of X-ray energies in which there is overlap with the platinum L_{III} EXAFS.

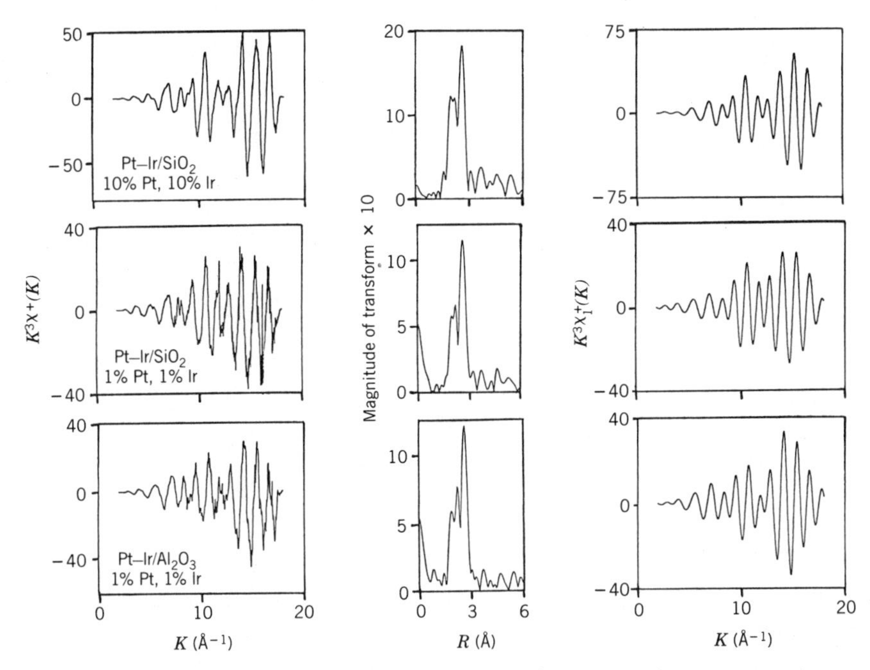

Figure 4.26 Normalized EXAFS data at 100°K, with associated Fourier transforms (and inverse transforms taken over the range of distances 1.4–3.3 Å), for the extended fine structure beyond the platinum L_{III} edge for a series of platinum–iridium catalysts *(48)*. (Reprinted with permission from the American Institute of Physics.)

We begin by considering the iridium EXAFS of a reference material such as metallic iridium or a catalyst containing pure iridium clusters. An EXAFS function for the iridium in the platinum–iridium catalyst is then generated from the function for the reference material by introducing adjustments for differences in interatomic distances, amplitude functions, and phase shifts. In making such adjustments, we are aided by the fact that the amplitude functions and phase shift functions of platinum are not very different from those of iridium, as shown in Figures 4.27 and 4.28.

The amplitude functions in Figure 4.27 were determined from EXAFS data on pure metallic platinum and iridium reference materials *(48)*. The amplitude functions are modified by multiplication by the factor (R^2_1/N_1), where R_1 and N_1 are the nearest neighbor distance and coordination number, respectively. Both the shape and magnitude of the amplitude functions are very similar.

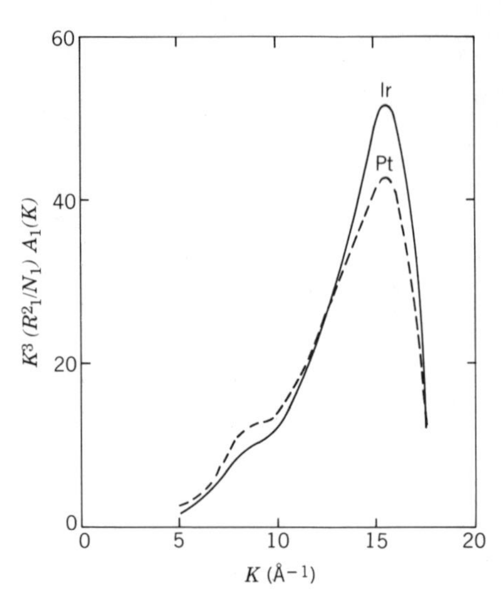

Figure 4.27 Modified amplitude functions (for the first coordination shell of atoms) for the EXAFS associated with the L_{III} absorption edges of pure metallic platinum and iridium at 100°K *(48)*. (Reprinted with permission from the American Institute of Physics.)

Because of the similarity in the amplitude functions of platinum and iridium, we do not separate the backscattering contributions of platinum and iridium atoms in the analysis of EXAFS data on platinum-iridium clusters or alloys. In our quantitative treatment of EXAFS arising from nearest neighbor atoms of platinum and iridium, the EXAFS function of Eq. 4.3 consists of only one term, as will be seen in the following discussion.

To obtain structural information on platinum–iridium clusters from EXAFS data, we concentrate primarily on the determination of interatomic distances. To obtain accurate values of interatomic distances, we need to have precise information on phase shifts. In this regard, we are fortunate that the phase shift functions of platinum and iridium are not very different.

In Figure 4.28 phase shift functions are shown for the various possible combinations of absorber and backscattering atoms in the platinum–iridium system *(48)*. For the platinum–iridium pair there are two functions, since there are two different combinations of absorber and backscattering atoms. The two functions are distinguished by using the designation PtIr when Pt is the absorber atom and IrPt when Ir is the absorber atom.

The phase shift functions for PtPt and IrIr were determined from data on

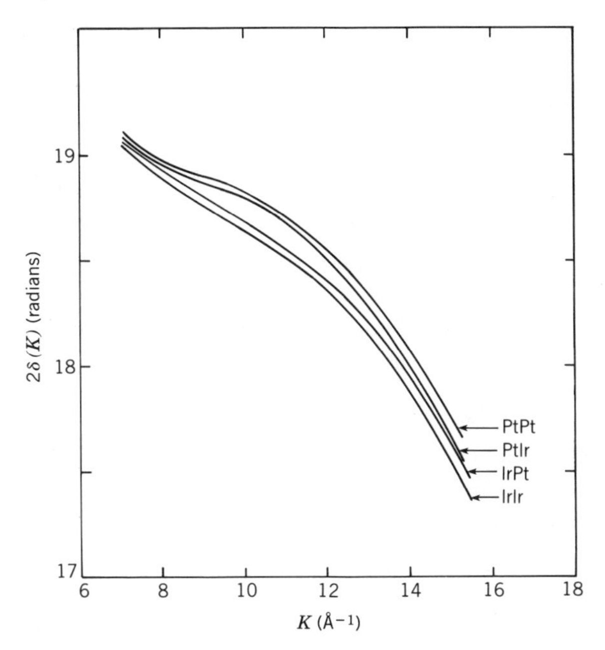

Figure 4.28 Phase shift functions for platinum–platinum, iridium–iridium, and platinum–iridium atomic pairs *(48)*. (For the platinum–iridium pair there are two functions, designated PtIr when platinum is the absorber atom and IrPt when iridium is the absorber atom.) (Reprinted with permission from the American Institute of Physics.)

pure platinum and iridium. Theoretical phase shifts from Teo and Lee *(34)* for PtPt and IrIr with an adjustment of only −3 eV in the absorption threshold energy agree very well with the experimental phase shifts. Since the experimentally derived phase shift functions for PtPt and IrIr in Figure 4.28 are not very different, the functions for PtIr and IrPt were obtained from them by making small adjustments based on the tabulated theoretical phase shifts of Teo and Lee for individual absorber and backscattering atoms.

The phase shifts of interest in the analysis of platinum L_{III} EXAFS for platinum–iridium catalysts are those for PtPt and PtIr. The difference in phase shifts for these combinations over the whole range of K (7 to 15 $Å^{-1}$) shown in Figure 4.28 is only 0.03 to 0.09 radian. If a platinum atom had an equal number of platinum and iridium atoms as nearest neighbors on the average, as in a completely random distribution of the two types of atoms in a material containing equal atomic fractions of platinum and iridium, the average phase shift would differ from that of pure platinum by only 0.015 to 0.045 radian over the range of K shown.

Similar considerations apply to the phase shifts of interest in the analysis of the iridium L_{III} EXAFS for platinum–iridium catalysts. Therefore, for simplicity, the phase shift functions for PtPt and IrIr are used in the analysis of the EXAFS associated with the platinum and iridium edges, respectively. This simplifying assumption introduces an uncertainty of only about 0.001 Å in the interatomic distances derived from the data.

Returning to our discussion of the contribution of the function $I_1(K)$ to the extended fine structure associated with the platinum L_{III} edge of a platinum–iridium catalyst, we generate an EXAFS function for the iridium in the catalyst using the amplitude function $A^R(K_{Ir})$ for an iridium reference material, either pure iridium or an iridium reference catalyst *(48).* The superscript R signifies that the amplitude function is for the reference material while the subscript Ir on K denotes that the wave vector is defined for iridium EXAFS.

The iridium EXAFS function for the platinum–iridium catalyst may be written as follows:

$$\chi(K_{Ir}) = A^R(K_{Ir}) \times C \exp(-2K_{Ir}^2\Delta\sigma^2) \times \sin[2K_{Ir}R + 2\delta(K_{Ir})] \qquad (4.14)$$

where C is a scaling factor related to coordination numbers and $\Delta\sigma^2$ is the difference between the value of σ^2 characteristic of the iridium L_{III} EXAFS of the platinum–iridium catalyst and the corresponding value of σ^2 for the iridium reference material. Although the subscript 1 has been deleted from a number of the symbols in Eq. 4.14 for simplicity, the equation applies only to the contribution of nearest neighbor atoms to the EXAFS. The phase shift $2\delta(K_{Ir})$ is taken to be equal to that for pure iridium for reasons we have already discussed.

To obtain the desired function $I_1(K)$ from Eq. 4.14, we require a relation between the wave vector K calculated from the threshold energy E_0^{Pt} associated with the platinum L_{III} edge and the wave vector K_{Ir} calculated from the threshold energy E_0^{Ir} associated with the iridium L_{III} edge *(48).* The required relation is

$$K^2 = K_{Ir}^2 - \frac{2m}{\hbar^2}(E_0^{Pt} - E_0^{Ir}) \qquad (4.15)$$

where m and $\hbar$ have been defined in Eq. 4.1. For a given value of $\chi(K_{Ir})$ in Eq. 4.14 there will be a corresponding value of $I_1(K)$ in Eq. 4.13. The function $I_1(K)$ may be written in the following form:

$$I_1(K) = G(K) \times C \exp\left\{-2\Delta\sigma^2\left[K^2 + \frac{2m}{\hbar^2}(E_0^{\mathrm{Pt}} - E_0^{\mathrm{Ir}})\right]\right\} \quad (4.16)$$

where the function $G(K)$ is obtained simply by associating values of the wave vector K with values of the function

$$A^R(K_{\mathrm{Ir}}) \times \sin[2K_{\mathrm{Ir}}R + 2\delta(K_{\mathrm{Ir}})]$$

from Eq. 4.14.

To determine values of interatomic distances we return to Eq. 4.13. The platinum EXAFS function is given by the equation

$$\chi_1(K) = A^R(K) \times D \exp(-2K^2\Delta\sigma^2) \times \sin[2KR + 2\delta(K)] \quad (4.17)$$

The quantity D is a scaling factor related to coordination numbers, and $\Delta\sigma^2$ is the difference between the value of σ^2 characteristic of the platinum L_{III} EXAFS of the platinum–iridium catalyst and the corresponding value of σ^2 for a platinum reference material *(48)*. The reference material is pure metallic platinum or a platinum catalyst. The quantity $A^R(K)$ is the amplitude function of the platinum reference material. It is emphasized that the parameter $\Delta\sigma^2$ in Eq. 4.17 is different from the analogous parameter in Eq. 4.14. Also, the values of the interatomic distance R in Eqs. 4.14 and 4.17 will not in general be equivalent.

The determination of interatomic distances and other parameters involves fitting of the expression for $\chi_1^+(K)$ in Eq. 4.13, utilizing Eqs. 4.16 and 4.17 for the terms $I_1(K)$ and $\chi_1(K)$, respectively, to the corresponding function derived from the experimental EXAFS data. The fitting is accomplished with the least squares procedure referred to earlier. In the application of the fitting procedure, values of the parameters C and $\Delta\sigma^2$ of Eq. 4.14 and of the parameters D, $\Delta\sigma^2$, and R of Eq. 4.17 are determined for various values of R in Eq. 4.14, the latter being designated by R_{Ir} in the following discussion. The quality of fit, as represented by a fitting error or standard deviation of fit of the expression for $\chi_1^+(K)$ of Eq. 4.13 to the corresponding quantity derived from experiment, is determined as a function of R_{Ir}. The value of R_{Ir} at which the fitting error has its minimum value is the "best fit" value for the characterization of the function $I_1(K)$. The interatomic distance R of Eq. 4.17 obtained for a particular value of R_{Ir} is designated as R_{Pt} in the following discussion. The value of R_{Pt} corresponding to the minimum fitting error is taken as the "best fit" value of the interatomic distance characterizing the platinum L_{III}

EXAFS, that is, the distance between a platinum absorber atom and nearest neighbor atoms of platinum or iridium.

The method of accounting for the overlapping of iridium L_{III} EXAFS in the analysis of data on the fine structure associated with the platinum L_{III} edge of platinum–iridium catalysts has been tested in EXAFS experiments on a physical mixture of platinum and iridium and on a bulk alloy of the two elements *(48).* Figure 4.29 shows the effect of the value assumed for R_{Ir} on the quality of fit to the experimental data for $\chi_1^+(K)$ that can be obtained by Eqs. 4.13, 4.16, and 4.17. The quality of fit is characterized by the normalized fitting error, which is defined in this case as the ratio of the standard deviation of fit for a given value of R_{Ir} to the minimum standard deviation obtained over the range of distances examined.

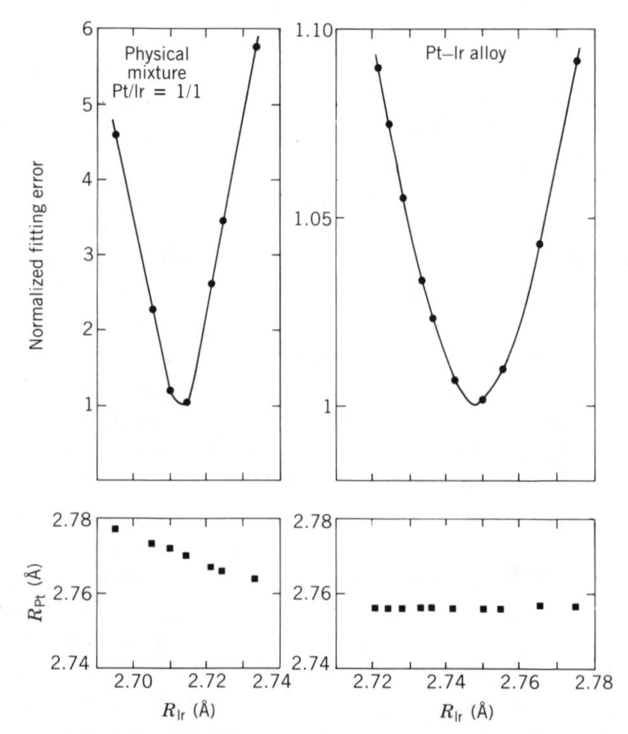

Figure 4.29 Test of method of accounting for the overlapping of iridium L_{III} EXAFS in the analysis of platinum L_{III} EXAFS data on a physical mixture of platinum and iridium and on a platinum–iridium bulk alloy *(48).* (Reprinted with permission from the American Institute of Physics.)

The two upper sections of Figure 4.29 show the normalized fitting error as a function of R_{Ir} for the physical mixture and the alloy. The two lower sections show the values obtained for the distance R_{Pt} for various values for R_{Ir}. The left-hand sections of the figure refer to the physical mixture, which was diluted with alumina to give a sample containing 10 wt% each of platinum and iridium. The right-hand sections refer to the bulk alloy, which contained 58 wt% platinum and 42 wt% iridium.

For the physical mixture, the values of R_{Ir} and R_{Pt} for the best fit to the data would be expected to be the known interatomic distances for the pure metals, 2.714 and 2.775 Å *(33),* respectively. In Figure 4.29 the normalized fitting error passes through a minimum at R_{Ir} equal to 2.713Å, which differs by only 0.001 Å from the known value for pure metallic iridium. In the lower left-hand section of the figure the value of R_{Pt} corresponding to the minimum fitting error is 2.771 Å, which is 0.004 Å lower than the known value of 2.775 Å. These results indicate that the method of accounting for the overlapping of iridium L_{III} EXAFS is satisfactory and that distances can be determined with an uncertainty of about 0.005 Å or lower.

In the application of the method to the bulk alloy of platinum and iridium, the values of the distances R_{Ir} and R_{Pt} would be expected to be equal at the minimum of a plot of fitting error vs. R_{Ir} if the alloy were completely homogeneous. Inspection of the right-hand sections of Figure 4.29 shows that the distances differ by only 0.008 Å, so that either distance differs from the average by only 0.004 Å. This difference is in the range of our estimated uncertainty. If the difference between the values of 2.748 and 2.756 Å for R_{Ir} and R_{Pt}, respectively, were significant, the results would indicate that the alloy is not completely homogeneous, that is, an atom of platinum or iridium on the average would be surrounded by a greater number of nearest neighbor atoms of its own kind than would be expected for a completely random distribution.

The interatomic distance determined from X-ray diffraction data on the alloy is 2.751 Å, which is very close to the average of the two distances derived from Figure 4.29. If one assumes a linear relation between interatomic distance and alloy composition (Vegard's law), the value of 2.751 Å would correspond to an alloy composition of 60% platinum, 40% iridium. The alloy composition as determined by X-ray fluorescence is 58% platinum, 42% iridium *(48).*

When the method is applied to the EXAFS data on platinum–iridium catalysts given in Figure 4.26, the results shown in Figure 4.30 are obtained *(48).* For the silica-supported platinum–iridium catalyst containing 10 wt% each of platinum and iridium, the values of R_{Pt} and R_{Ir} corresponding to the minimum fitting error are 2.753 and 2.724 Å, respectively. This difference is well beyond our estimated uncertainty of ± 0.005 Å in distance determination. The 2.753 Å distance is significantly lower than the distances of 2.775 and 2.774 Å, respectively, for pure platinum and for the platinum clusters in a Pt/SiO_2 reference catalyst. The 2.724-Å distance is significantly higher than the distances of 2.714 and 2.712 Å, respectively, for pure iridium and for the iridium clusters in an Ir/SiO_2 reference catalyst *(48).* The distances in the metal clusters in the reference catalysts were obtained from EXAFS data.

The fact that different distances are obtained for R_{Pt} and R_{Ir} in the platinum–iridium catalyst indicates that the average composition of the first coordination shell of atoms surrounding a platinum atom is different from that surrounding an iridium atom. The data suggest that platinum-rich and iridium-rich regions exist in the catalyst and could be taken as evidence for the presence of metal clusters with different compositions, some of which are platinum-rich and others of which are iridium-rich.

Alternatively, one can visualize platinum-rich and iridium-rich regions

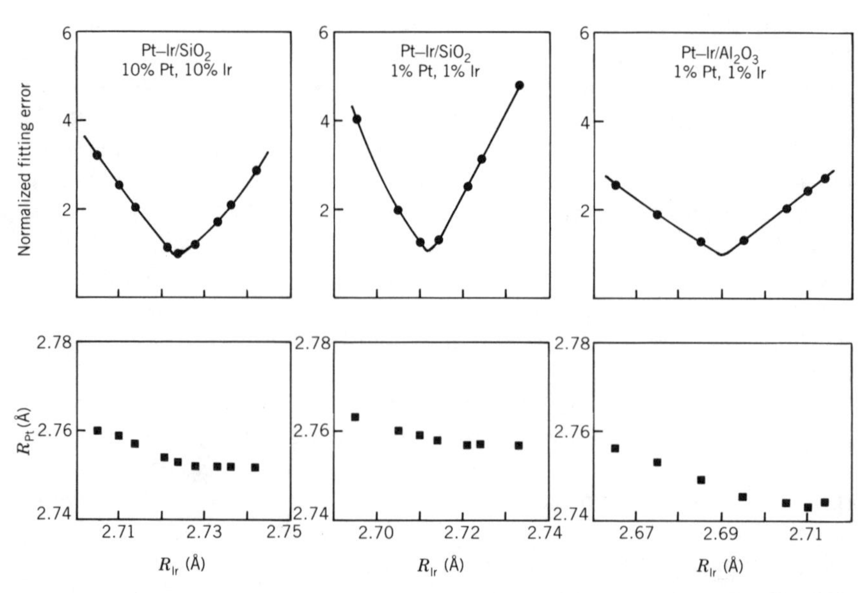

Figure 4.30 Determination of interatomic distances in platinum–iridium catalysts by evaluation of the quality of fit of the expression for the sum of the platinum L_{III} EXAFS function and the function $I_1(K)$ arising from the iridium L_{III} EXAFS to the corresponding sum derived from experimental data *(48).* (Reprinted with permission from the American Institute of Physics.)

within a given metal cluster. From surface energy considerations, it might be anticipated that the platinum-rich region would be present at the surface *(48)*. Platinum, which has a lower heat of sublimation than iridium *(49),* would be expected to have the lower surface energy *(50)*. The iridium-rich region would then constitute the central core of the cluster, so that a composition gradient would exist from the inside to the outside of the cluster. The detailed nature of such a concentration gradient would have to be consistent with the symmetrical lines observed in the X-ray diffraction pattern on the catalyst. Experimental evidence for the concentration of platinum in the surface has been obtained recently from studies on platinum–iridium films *(51)*.

While the X-ray diffraction data considered earlier for the catalyst containing 10 wt% each of platinum and iridium do not eliminate the possibility of differences in the average environments about the two types of atoms in the clusters, they also do not provide evidence for it. The diffraction data are entirely consistent with the platinum and iridium being present as homogeneous bimetallic clusters. This interpretation is reasonable in the absence of other information. The results of the EXAFS studies, however, provide evidence that the catalyst consists of platinum–iridium clusters that are not homogeneous. Hence the EXAFS data provide more detailed structural information than can be obtained from a diffraction pattern.

When the platinum–iridium clusters on silica are more highly dispersed, as in the case of the catalyst containing 1 wt% each of platinum and iridium, the results shown in the middle sections of Figure 4.30 are obtained *(48)*. The metal dispersion (ratio of surface atoms to total metal atoms) of the clusters as determined by hydrogen chemisorption is 0.75. Assuming spherically shaped clusters, an average diameter of about 15 Å is obtained from the chemisorption data. The clusters in this case are too small to give a satisfactory X-ray diffraction pattern. Figure 4.30 shows that the values of R_{Pt} and R_{Ir} corresponding to the minimum fitting error are different, 2.758 and 2.712 Å, respectively. The 2.758-Å distance is significantly different from the interatomic distance of 2.774 Å for the platinum clusters in a Pt/SiO_2 reference catalyst, but the 2.712-Å distance is indistinguishable from the distance determined for the iridium clusters in an Ir/SiO_2 reference catalyst.

When the platinum–iridium clusters are still more highly dispersed, and are supported on an alumina carrier instead of silica, the results shown in the right-hand sections of Figure 4.30 are obtained *(48)*. The metal dispersion of the clusters as determined by hydrogen chemisorption is 0.93. If the clusters were spherical, the average diameter calculated from the chemisorption data would be about 12 Å. Again the clusters are too small to give a satisfactory X-ray diffraction pattern. As with the previous two catalysts, the values of R_{Pt} and R_{Ir} corresponding to the minimum fitting error are different, 2.747 Å and 2.690 Å, respectively. For this catalyst, however, there is the new feature that *both* distances are significantly lower than those, 2.758 Å and 2.704 Å, respectively, which we obtain for the pure metal clusters in Pt/Al_2O_3 and Ir/Al_2O_3 reference catalysts from EXAFS data *(48).*

In contrast to the metal clusters in the Pt/SiO_2 and Ir/SiO_2 reference catalysts *(19),* those in the Pt/Al_2O_3 and Ir/Al_2O_3 reference catalysts exhibit interatomic distances lower than the distances in the corresponding pure metals, which are 2.775 Å and 2.714 Å *(33),* respectively, for platinum and iridium. The contraction observed when the clusters are dispersed on alumina indicates an interaction with the carrier that is not apparent in the silica-supported clusters. The finding that the distance contractions are more pronounced for the bimetallic platinum–iridium catalyst than for the monometallic reference catalysts provides additional evidence that the bimetallic catalyst is not simply a mixture of platinum clusters and iridium clusters.

By applying the method described here to account for the overlap of the iridium and platinum EXAFS in the extended fine structure associated with the L_{III} edge of platinum, it has been possible to obtain information on interatomic distances in a series of platinum–iridium catalysts. The ability of Eq. 4.13 for $\chi_1^+(K)$ to fit the corresponding function derived from the experimental data, with the use of Eqs. 4.16 and 4.17 for the functions $I_1(K)$ and $\chi_1(K)$, respectively, is illustrated in Figure 4.31 for the various catalysts *(48).*

For each catalyst, the experimental function $\chi_1^+(K)$ multiplied by K^3 is represented by the solid line which is repeated in the three sections labeled *a, b,* and *c.* The dotted line in each section labeled *a* represents the product of K^3 and the function $\chi_1^+(K)$ calculated from Eqs. 4.13, 4.16, and 4.17 using the parameters corresponding to the minimum fitting error in Figure 4.30. In general, the quality of fit is very good. In sections *b* and *c,* respectively, the dotted lines show the individual contributions of the platinum L_{III} EXAFS and the function $I_1(K)$.

In view of the expectation that platinum will concentrate in the surface of platinum–iridium clusters, we might anticipate that platinum and iridium would segregate from one another to an increasingly greater extent as the clusters become smaller and the ratio of surface atoms to total metal atoms increases *(48).* When the ratio is equal to 0.5 for clusters containing 50% each of platinum and iridium, one can visualize a situation in which essentially all of the platinum is present in the surface and all of the iridium in the interior. There would then be a close resemblance to the ruthenium–copper clusters *(2,31)* discussed earlier.

When the ratio of surface atoms to total atoms in the clusters is increased

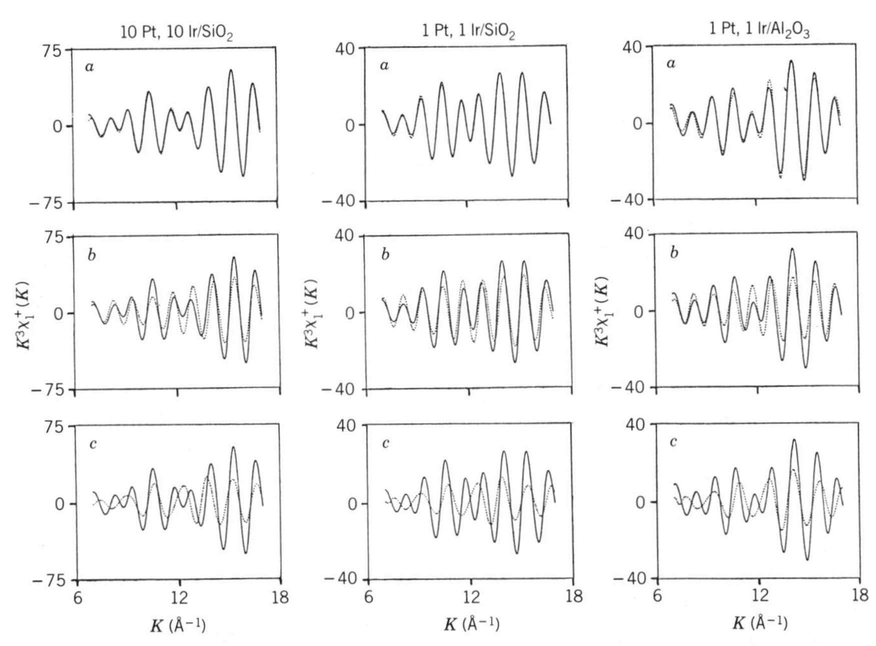

Figure 4.31 Contributions of the platinum L_{III} EXAFS, represented by the dotted lines in the fields labeled *b,* and the function $I_1(K)$, represented by the dotted lines in the fields labeled *c,* to the extended fine structure (solid lines) associated with the platinum L_{III} edges of platinum–iridium catalysts *(48).* [In each field labeled *a* the dotted line was determined by fitting the expression for the sum of the platinum EXAFS and the function $I_1(K)$ to the solid line, which was obtained from Figure 4.26.] (Reprinted with permission from the American Institute of Physics.)

beyond 0.5 for a composition of 50% platinum, 50% iridium, there is not enough platinum present to occupy the entire surface. When such a high degree of metal dispersion is attained, the metal clusters become extremely small (smaller than about 20 Å in diameter if they are spherical).

When the ratio of surface atoms to total atoms approaches unity, the notion of complete or nearly complete segregation of the platinum in a surface layer and of iridium in a central core cannot be accommodated if the clusters are spherically symmetrical. The notion can, however, be accommodated without difficulty if the clusters have a two-dimensional, "raftlike" shape rather than a spherical shape *(48).* One could then visualize a central iridium or iridium-rich raft with platinum atoms around the perimeter. In very highly dispersed catalysts of this type, the effect of the platinum on the catalytic properties of the iridium, and vice versa, would presumably be a consequence of the interaction between the two components at the boundary.

Mössbauer Effect Spectroscopy Studies (41). Another physical probe which has been used in the characterization of platinum–iridium catalysts is Mössbauer effect spectroscopy *(3,41,52)*. It can be applied to catalysts in which virtually all of the metal atoms are surface atoms. In Mössbauer effect spectroscopy one is concerned with a transition between a ground state and an excited state of a nucleus *(53)*.

As a specific example, let us consider a transition of an ^{57}Fe nucleus involving the emission or absorption of a gamma ray. Assume that the ^{57}Fe nucleus is present in a solid lattice such as a metal crystallite in a catalyst. If the energy of a gamma ray emitted by some source is exactly equal to the transition energy of the ^{57}Fe nucleus in the crystallite, the condition for resonance absorption is satisfied. Since the ^{57}Fe nucleus is fixed in a solid, it can absorb a gamma ray photon without recoil. This situation differs from that of a free nucleus in that the recoil momentum is imparted to the solid as a whole. Since the energy which goes into the motion of the solid as a whole is negligible, the full energy of the gamma ray photon is available for the transition.

The energy of the transition will vary slightly with the environment of the ^{57}Fe nucleus. To make it possible to observe the transitions in different types of environments with a given source of gamma rays, the source is moved with respect to the absorber. This movement results in a Doppler shift in the energy of the gamma ray. A given change in the transition energy in the absorber material will correspond to a particular velocity of the source relative to the absorber. The velocity corresponding to a given variation in transition energy is called the "isomer shift." By measuring the isomer shift of the ^{57}Fe nucleus in a given sample, information on the chemical state of the iron atom, or on the electronic environment of the ^{57}Fe nucleus, can be obtained.

Another parameter of interest is the quadrupole splitting, which arises as the result of the presence of an electric field gradient at the ^{57}Fe nucleus. The interaction of the electric field gradient with the nuclear quadrupole moment splits the excited state of spin 3/2 into two states, so that two resonances may be observed. The quadrupole splitting is expressed as the difference in Doppler velocities corresponding to the two resonances.

In conducting Mössbauer effect spectroscopy experiments on platinum–iridium catalysts, one might incorporate the Mössbauer nuclides ^{195}Pt and/or ^{193}Ir in the catalysts. However, experiments with these nuclides are more difficult because of short-lived sources and the requirement for measurements at liquid helium temperature. An alternative approach involves the addition of ^{57}Fe to platinum–iridium catalysts as a sensitive Mössbauer probe of the interaction between platinum and iridium. Since the Mössbauer resonance is sensitive to the environment of the ^{57}Fe nuclei, it is possible to obtain information on the nature of the catalyst.

Mössbauer spectra at 25°C are shown in Figure 4.32 for alumina-supported platinum, iridium, and bimetallic platinum–iridium catalysts containing ^{57}Fe (samples B, C, and D, respectively) *(3,41)*. The platinum–iridium catalyst contained 1.75 wt% each of platinum and iridium, while the other two catalysts contained 1.75 wt% of either platinum or iridium. All of the catalysts had metal dispersions (as determined by chemisorption) in the range 0.7–1. Also

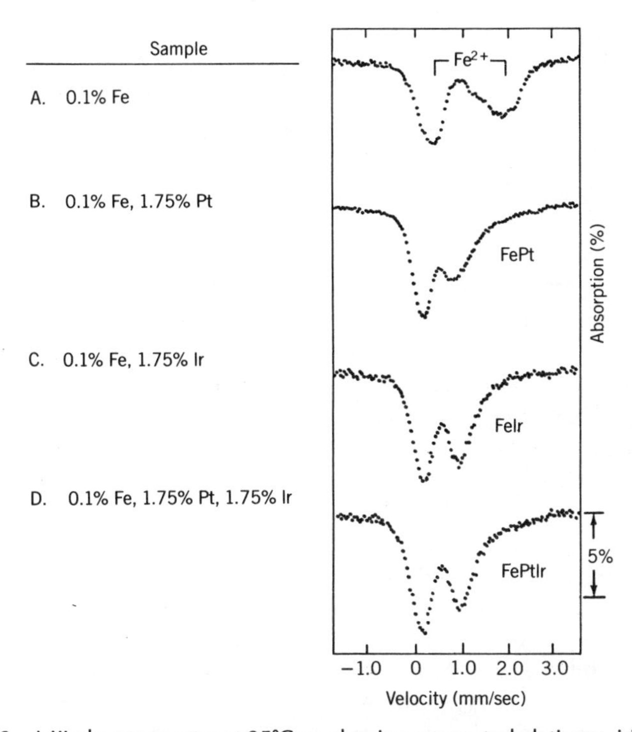

Figure 4.32 Mössbauer spectra at 25°C on alumina-supported platinum, iridium, and platinum–iridium catalysts (samples B, C, and D, respectively) containing a small amount of Fe (0.1 wt%) enriched with ^{57}Fe as a Mössbauer probe *(3,41)*. (Sample A is a reference material containing only the enriched Fe on alumina.) (Reprinted with permission from Academic Press, Inc.)

shown is a reference spectrum for alumina containing only the iron (sample A). A spectrum (not shown in Figure 4.32) was also obtained on a physical mixture of samples B and C designated as sample B-C.

The two-line (quadrupole split) spectrum for sample A and the corresponding Mössbauer parameters for this sample in Table 4.2 are characteristic of high spin ferrous ions in oxides *(54),* which indicates that the iron in the sample is reduced only to the ferrous state with hydrogen at 500°C. There is no evidence of iron in the metallic state. The large linewidths and the differences in the intensities of the lines suggest that the iron has a number of different environments.

The spectra for samples B, C, and D in Figure 4.32 are very different from that for sample A. Correspondingly, the Mössbauer parameters in Table 4.2 for samples B, C, D, and B-C contrast markedly with those for sample A. These results are indicative of differences in the chemical environment of the iron. The presence of platinum and/or iridium leads to a marked decrease in the amount of iron which exists as ferrous ions in the samples. The iron is therefore present in a different chemical state, which is characterized by an intimate association with highly dispersed clusters of platinum and/or iridium.

The evidence for association of the iron atoms with platinum in samples similar to B has been discussed elsewhere in some detail *(55).* Incorporation of iron atoms into the platinum clusters is indicated by the agreement of the isomer shift for iron in the catalyst with that for iron in a PtFe bulk alloy of similar composition. In addition, it is found that the iron in samples such as B, after reduction at 500°C, exhibits a unique chemical behavior in that it can be reversibly oxidized and reduced at room temperature. This property appears to be characteristic of iron in association with metals such as platinum and palladium *(54–57).* It is not observed when the iron is present by itself on alumina.

Sample C, containing iridium and iron, exhibits the same reversible oxidation–reduction behavior, indicating the incorporation of the iron into the iridium clusters. The isomer shift for sample C, however, was not in good agreement with the value of 0.38 mm sec^{-1} *(58)* expected for dilute iron in iridium alloys. This difference may reflect an unusual chemical state for iron which is associated with surface iridium atoms in clusters. A similar situation has recently been reported and discussed for iron–ruthenium catalysts *(59).*

The Mössbauer parameters for sample D differ from those for samples B and C, indicating that the iron atoms in sample D are not associated exclusively with either the platinum or the iridium in the catalyst. Furthermore, the area and linewidth ratios (A_2/A_1 and W_2/W_1) for sample D differ from

Table 4.2 Summary of Mössbauer Parameters for Reduced Pt, Ir, and PtIr Catalysts Containing ^{57}Fe *(41)*

Composition[a], wt%	Sample[b]	Isomer Shift δ, mm/sec	Quadrupole Splitting, mm/sec	A_2/A_1[c]	W_2/W_1[c]
0.1 Fe	A	1.20	1.69	1.01	1.32
0.1 Fe, 1.75 Pt	B	0.53	0.82	1.71	2.41
	B-600	0.55	0.86	1.90	2.73
	$B\text{-}H_2O$[d]	0.53	0.83	1.71	2.38
0.1 Fe, 1.75 Ir	C	0.58	0.97	1.18	1.58
	C-600	1.22	1.86	0.98	1.38
0.1 Fe, 1.75 Pt, 1.75 Ir	D	0.56	0.89	1.16	1.38
	D-600	0.57	0.92	1.50	2.16
	B-Ir[d]	0.57	0.86	1.13	1.40
0.1 Fe, 1.75 Pt mixed with 0.1 Fe, 1.75 Ir	B-C[e]	0.57	0.92	1.28	1.79

NOTE: Reprinted with permission from Academic Press.

[a]The concentrations of metallic components of interest, expressed as wt% of the total mass of material, which includes the metallic elements shown and the alumina carrier. The ^{57}Fe fraction of the 0.1 wt% Fe present in all cases serves as a Mössbauer probe of the catalyst. The composition listed as 0.1 Fe (sample A) is a reference material in which Fe is dispersed on alumina with no platinum or iridium present.

[b]Samples A, B, $B\text{-}H_2O$, C, D, and B-Ir were all calcined in air at 260°C in their preparation, while samples B-600, C-600, and D-600 were calcined at 600°C.

[c]The ratios of the areas (A_2/A_1) and widths (W_2/W_1) of the two lines in the Mössbauer spectra in Figures 4.32, 4.34, and 4.35. The subscript 1 refers to the line corresponding to the lower Doppler velocity.

[d]Catalyst samples prepared by contacting sample B in its finally reduced form with water (designated $B\text{-}H_2O$) or with a chloroiridic acid solution (designated B-Ir), followed by drying and reduction steps identical to those employed in the preparation of the original sample B.

[e]Physical mixture of equal portions of samples B and C.

those for the physical mixture B-C, indicating that the iron atoms in sample D are not simply distributed between platinum clusters and iridium clusters with the platinum and iridium being totally isolated from each other. The data on Mössbauer parameters are consistent with the view that catalyst D consists of bimetallic platinum–iridium clusters with iron atoms incorporated therein.

Mössbauer spectra obtained at very low temperatures (18–23.5°K) on samples B, C, and D are shown in Figure 4.33. At these low temperatures the samples become ferromagnetic, and the spectra exhibit magnetic hyperfine splitting *(3,41)*. As a result, the spectra consist of six lines instead of two. While the six lines in the spectra of Figure 4.33 are not very pronounced, they are nonetheless real. The positions of the lines, as determined from a computer analysis of the data, are shown above the spectra in Figure 4.33.

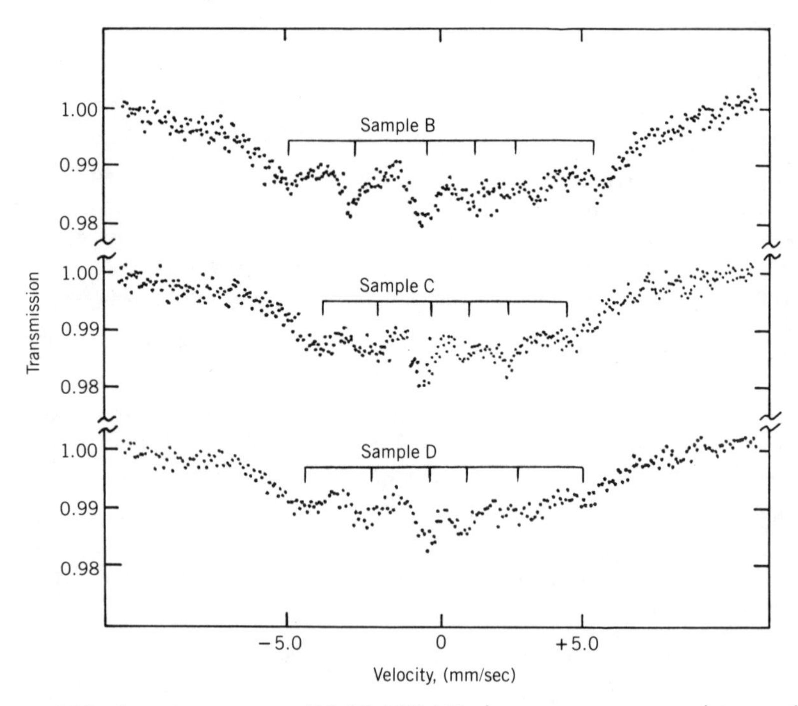

Figure 4.33 Low temperature (18–23.5°K) Mössbauer spectroscopy data on the same catalysts (samples B, C, D) used in obtaining the room temperature Mössbauer spectroscopy data in Figure 4.32 *(3,41)*. (Reprinted with permission from Academic Press, Inc.)

The magnitude of the magnetic field experienced by the ^{57}Fe nuclei can be determined from the separations of lines in the spectrum *(53,60)*. Magnetic fields derived from the separation of the outermost of the six lines of the hyperfine spectra in Figure 4.33 are given in Table 4.3. Since the internal magnetic field is different for each of the three samples, which differ with regard to the metal or metals (platinum and/or iridium) present, the results provide additional evidence that the iron atoms are associated with these metals.

In studies on bulk PtFe, IrFe, and PtIrFe alloys, it was observed *(61,62)* that the magnetic fields at the iron nuclei were much smaller for IrFe than for PtFe. These studies showed also that the magnetic fields for PtIrFe alloys were intermediate between those for PtFe and IrFe alloys. The same trend is found with samples B, C, D in Table 4.3, which is consistent with the view that the iron atoms in sample D are incorporated in bimetallic clusters of platinum and iridium.

The magnetic field at the ^{57}Fe nuclei in the platinum catalyst (sample B) is similar to that reported for highly dispersed PtFe alloys on carbon *(57)* and for bulk PtFe alloys *(62)*. However, the fields at the ^{57}Fe nuclei in the iridium and platinum–iridium catalysts (samples C and D) are much larger than those observed with bulk IrFe and PtIrFe alloys *(61,62)*. The data thus indicate a strong effect of the degree of metal dispersion on the magnetic properties of the IrFe and PtIrFe systems that is not observed with the PtFe system.

In the preparation of iridium-containing catalysts, the temperature of calcination in air is critical. Mössbauer spectra at 25°C are shown in Figure 4.34 for platinum, iridium, and bimetallic platinum–iridium catalysts (samples B-600, C-600, and D-600, respectively) which were prepared by calcination in air at 600°C prior to the final reduction in hydrogen at 500°C. The catalysts have the same elemental compositions as the catalysts for which spectra are shown in Figure 4.32, but they exhibit structural differences due to the higher calcination temperature (600 vs. 260°C) employed in their preparation.

Table 4.3 Magnetic Fields Derived from Magnetic Hyperfine Structure *(41)*

Sample and Composition, wt%	Temperature, °K	Magnetic Field, kOe
B (0.1 Fe, 1.75 Pt)	21	325
C (0.1 Fe, 1.75 Ir)	23.5	260
D (0.1 Fe, 1.75 Pt, 1.75 Ir)	18	295

NOTE: Reprinted with permission from Academic Press.

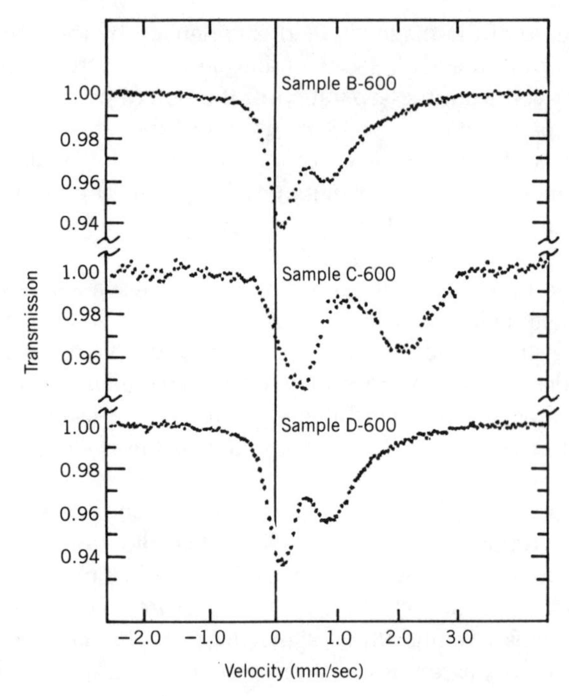

Figure 4.34 Mössbauer spectra at 25°C on alumina-supported platinum, iridium, and platinum–iridium catalysts which were calcined in air at 600°C in their preparation (samples B-600, C-600, and D-600, respectively) *(3,41).* (Reprinted with permission from Academic Press, Inc.)

Sample B-600 gives a spectrum similar to that of sample B in Figure 4.32, and the Mössbauer parameters for these samples in Table 4.2 are also similar. These results indicate that sample B-600 also consists of platinum clusters with iron atoms incorporated therein.

However, sample C-600 gives a spectrum which is very different from that presented for sample C in Figure 4.32. The spectrum for sample C-600 is nearly identical to that of sample A in Figure 4.32, in which the iron exists in the ferrous state. The iridium in sample C-600 is not present as highly dispersed clusters, as shown also by chemisorption data *(41).* It appears that the poorly dispersed iridium crystallites (dispersion <0.1) are not associated to a significant extent with the iron in the sample. Hence sample C-600 behaves like a sample containing no iridium. The Mössbauer parameters in Table 4.2 are consistent with this statement.

The spectrum of sample D-600 resembles that of sample B in Figure 4.32 more closely than it resembles the spectrum of sample D. The Mössbauer parameters in Table 4.2, especially the ratios A_2/A_1 and W_2/W_1, show substantial differences between samples D and D-600. The iridium in sample D-600 is largely present in the same poorly dispersed form as the iridium in sample C-600, as has been found from X-ray diffraction studies of similar samples. Sample D-600 therefore can be characterized approximately as consisting of highly dispersed platinum clusters incorporating iron atoms and separate iridium crystallites of much lower dispersion that are not significantly associated with iron atoms. This characterization is consistent with chemisorption data *(41)*.

The Mössbauer data on sample D-600, coupled with chemisorption data on this sample, show that calcination of alumina-supported platinum–iridium in air at 600°C is unsatisfactory for the formation of highly dispersed bimetallic clusters of platinum and iridium.

Mössbauer spectroscopy data on an alumina-supported platinum–iridium sample prepared by a sequential impregnation procedure are of interest in providing information on the tendency of platinum and iridium to form bimetallic clusters. The sample was prepared by contacting the platinum–alumina catalyst containing ^{57}Fe (sample B) with a chloroiridic acid solution after the hydrogen reduction step. On the basis of the evidence presented earlier, the sample, prior to contact with the chloroiridic acid solution, may be characterized as containing highly dispersed platinum clusters with iron atoms incorporated therein.

After drying and reduction steps identical to those employed in the preparation of samples B, C, and D, the sequentially impregnated material (sample B-Ir) had the same elemental composition as sample D. Another sample prepared in exactly the same manner, except that water was substituted for the chloroiridic acid solution, served as a reference material (designated as sample B-H_2O).

Mössbauer spectra at 25°C on these samples are presented in Figure 4.35. The spectrum and Mössbauer parameters of sample B-H_2O are similar to those of sample B (Figure 4.32 and Table 4.2). Sample B is therefore unaffected by its treatment with water. The spectrum of sample B-Ir is similar to that of sample D in Figure 4.32, and the Mössbauer parameters given in Table 4.2 are also similar.

The data indicate that platinum clusters associated with iron are altered by addition of iridium to the sample. We conclude that the added iridium is incorporated in the platinum clusters to give PtIr clusters containing the iron probe atoms. If the added iridium was present as separate iridium clusters

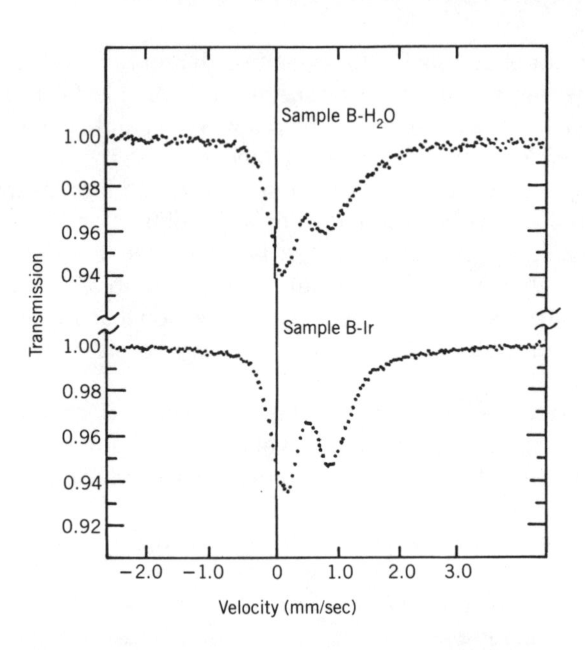

Figure 4.35 Mössbauer spectra at 25°C on catalyst samples B-H_2O and B-Ir prepared by contacting platinum–alumina catalyst (sample B of Figures 4.32 and 4.33) in its final reduced form with, respectively, water and a chloroiridic acid solution *(3,41)*. (The platinum and iridium contents of sample B-Ir are the same as those of sample D of Figure 4.32.) (Reprinted with permission from Academic Press, Inc.)

in sample B-Ir, the Mössbauer spectrum should have been the same as that of sample B-H_2O or sample B. These results provide a good illustration of the propensity of platinum and iridium atoms to mix and form bimetallic clusters, rather than to form monometallic clusters consisting exclusively of atoms of one or the other of the two metals.

4.3 PLATINUM–RHENIUM CATALYSTS

Another catalyst system that could be considered in the bimetallic cluster category is supported platinum–rhenium *(5),* which represents still another type of system in the sense that a Group VIIA metallic element (rhenium) is incorporated with the Group VIII metal component. Platinum and rhenium have different crystal structures (fcc vs. hcp) *(8)* and do not exhibit complete miscibility in the bulk (Ref. 45, p. 820). However, these factors may have limited significance for highly dispersed metals, as we have already seen in the ruthenium–copper and osmium–copper systems.

In considering the nature of platinum–rhenium catalysts, we begin with a comparison of the chemisorption properties of alumina-supported rhenium, platinum, and platinum–rhenium catalysts *(40)*. Data on the chemisorption of carbon monoxide and hydrogen at room temperature are given in Table 4.4 for catalysts with platinum and/or rhenium contents in the range of interest for reforming applications.

The data are expressed in terms of the quantities CO/M and H/M, which represent the number of adsorbed carbon monoxide molecules and hydrogen atoms, respectively, divided by the number of metal atoms (platinum and/or rhenium) in the catalyst. The values of CO/M represent the amounts of adsorbed carbon monoxide retained after the adsorption cell is evacuated at room temperature for 10 minutes to approximately 10^{-6} torr after completion

Table 4.4 Chemisorption Data on Platinum, Platinum–Rhenium, and Rhenium Catalysts *(40)*

Catalyst[a]	CO/M[b]	H/M[b]
0.3% Pt	0.8	1.0
0.3% Pt, 0.3% Re	0.7	0.4
0.3% Re	0.7	0.0

NOTE: Reprinted with permission from Elsevier Scientific Publishing Company.

[a] Alumina was the carrier in all of the catalysts. The catalysts were all calcined at 500–510°C in an oxygen–helium mixture containing 20% oxygen. They were then reduced in hydrogen at 500°C in adsorption cells, after which the cells were evacuated at 450°C and cooled to room temperature for the adsorption measurements.

[b] The quantities CO/M and H/M represent, respectively, the number of carbon monoxide molecules and hydrogen atoms chemisorbed at room temperature per atom of metal M (Pt and/or Re) dispersed on the alumina. The values were determined in the manner described in Table 4.1.

of the adsorption isotherm. The values of H/M were determined by extrapolation of the nearly pressure-independent regions of the adsorption isotherms to zero pressure.

All of the catalysts for which data are given in Table 4.4 were calcined (heated) in an oxygen–helium mixture containing 20% oxygen at 500–510°C. After the catalysts were charged to the adsorption cells, they were exposed to flowing hydrogen at 500°C in the manner normally employed for reducing the metallic component of the catalyst. The cells were then evacuated at 450°C and cooled to room temperature for determining adsorption isotherms.

The data on the catalyst containing rhenium alone indicate signficant chemisorption of carbon monoxide, but no chemisorption of hydrogen. As expected, the platinum catalyst chemisorbs both carbon monoxide and hydrogen, and the values of CO/M and H/M are nearly equal. The platinum–rhenium catalyst exhibits a value of CO/M about twice as high as the value of H/M. This result approximates what one would expect if hydrogen chemisorbed on only the platinum component of the catalyst. While this chemisorption behavior is consistent with the possibility that the platinum and rhenium are present as two separate entities in the catalyst, they do not rule out the possibility that bimetallic clusters of platinum and rhenium are present.

The physical and chemical nature of the rhenium in platinum–rhenium catalysts has been considered by a number of investigators. Johnson and Leroy *(63)* concluded that the rhenium is present as a highly dispersed oxide at typical reforming conditions. They studied a series of alumina-supported platinum–rhenium catalysts with platinum contents ranging from 0.31 to 0.66 wt% and rhenium contents ranging from 0.20 to 1.18 wt%. Their conclusions were based on measurements of hydrogen consumption during reduction of the catalysts at 482°C and on X-ray diffraction studies of the metal component of the catalyst after the alumina had been leached from the catalyst by treatment with a solution of fluoboric acid.

The catalysts were prepared by contacting alumina with aqueous solutions of chloroplatinic acid and ammonium perrhenate. The consumption of hydrogen during reduction corresponded to complete reduction of platinum from the +4 oxidation state to the metal and of rhenium from the +7 to the +4 state. The X-ray diffraction data on the metal residue from the leached catalysts showed no evidence for the presence of rhenium metal or a platinum–rhenium alloy. Most of the rhenium was found in the leaching solution. Finally, the authors stated that data from an electron spin resonance experiment on one of the reduced platinum–rhenium catalysts were consistent with their conclusion that the rhenium was present in the +4 state.

In contrast to the results of Johnson and Leroy, studies by Webb *(64)* on rhenium–alumina catalysts containing 3.42 and 3.64 wt% rhenium indicated that the rhenium was completely reduced from the +7 state to the metal by hydrogen at temperatures of 400–450°C. The catalysts were prepared by contacting alumina with an aqueous solution of perrhenic acid. Measurements of hydrogen consumption during reduction and of oxygen consumption during a subsequent oxidation were both consistent with changes of oxidation state between +7 and 0.

Two possible explanations were considered in accounting for the differences in results of these investigations. First, the authors *(64,65)* considered the possibility that the difference in the rhenium contents of the catalysts could play a role, since interaction of the rhenium with the alumina might be very strong at low concentrations and thereby inhibit reduction. Second, they noted that the water content of the hydrogen during reduction was lower in the experiments of Webb, due to his use of a system employing recirculating hydrogen from which water was continuously removed by means of a trap maintained at −196°C.

The experiments of Johnson and Leroy were conducted with a static system with product water remaining in the system, the water presumably being absorbed by the alumina in the catalyst. Johnson and Leroy estimated that the residual water pressure in their system was about 0.1 torr. It was implied *(65)* that the water level in the experiments of Johnson and Leroy was typical of that encountered in actual catalytic reforming reactors.

Later experiments by McNicol *(66)* and by a group of French workers *(67–72)*, in which the water formed during reduction was removed by a trap at 78°K, were consistent with the results of Webb in showing a change in oxidation state of rhenium from +7 to 0. These workers indicated also that the properties of alumina-supported platinum–rhenium catalysts depend on the method of preparation.

For catalysts that were simply dried in air at 110°C after impregnation of the alumina with H_2PtCl_6 and Re_2O_7, it was concluded that a platinum–rhenium alloy formed on reduction. This conclusion was based on the observation that the presence of platinum accelerated the reduction of oxygen chemisorbed on the rhenium and on results showing that the frequencies of the infrared absorption bands of carbon monoxide adsorbed on platinum and rhenium sites in platinum–rhenium catalysts were different from those found with catalysts containing only platinum or rhenium. However, for catalysts calcined in air at 500°C prior to reduction in hydrogen, it was concluded that the platinum exhibited much less interaction with the rhenium *(66,71)*.

Studies by a group at the Shell Laboratories in Amsterdam *(73)* have been reported as evidence of interaction between platinum and rhenium in catalysts containing these two elements. These workers, on the basis of infrared spectroscopy studies of carbon monoxide chemisorbed on platinum–rhenium catalysts, and also of X-ray photoelectron spectroscopy measurements on the catalysts, concluded that platinum–rhenium bonds are present in the surface.

In the infrared spectroscopy studies they observed that the stretching frequency of the chemisorbed carbon monoxide was higher on a $Pt–Re/SiO_2$ catalyst than on a Pt/SiO_2 catalyst. When they contacted a Re/SiO_2 catalyst with carbon monoxide, they observed only a weak band that rapidly disappeared and could not be restored on further exposure to carbon monoxide. They concluded that the carbon monoxide in this case dissociated, with resultant irreversible poisoning of the rhenium by carbon. Consequently, they attributed the infrared band observed for the platinum–rhenium catalyst to carbon monoxide chemisorbed on platinum atoms. It was concluded that the band was shifted in frequency from that observed for the platinum catalyst because of the interaction between platinum and rhenium.

In the X-ray photoelectron spectroscopy studies on the platinum–rhenium catalyst, it was observed that the binding energies of the platinum $4f_{7/2}$ and rhenium $4d_{5/2}$ core electrons were higher than they were in the catalysts containing platinum or rhenium alone.

The same directional changes in either CO stretching frequencies or core electron binding energies were observed when sulfur was chemisorbed on the platinum–rhenium catalyst. The effect of the sulfur was studied because it is commonly added to catalysts of this type in reforming practice, as will be pointed out in Chapter 5. It was concluded that the chemisorbed sulfur atoms are present on surface rhenium atoms, and that the latter remain bonded to platinum atoms in the surface region.

Thus the Shell workers conclude that Pt–Re bonds are present at the surface in platinum–rhenium catalysts, whether or not chemisorbed sulfur is present. A platinum–rhenium entity of the type envisioned by these investigators, if it indeed exists, could also be called a bimetallic cluster.

REFERENCES

1. Sinfelt, J.H. "Supported 'bimetallic cluster' catalysts." *J. Catal.* **29**: 308–315; 1973.

2. Sinfelt, J.H. "Catalysis by alloys and bimetallic clusters." *Acc. Chem. Res.* **10:** 15–20; 1977.

3. Sinfelt, J.H., inventor; Exxon Research and Engineering Company, assignee. "Polymetallic cluster compositions useful as hydrocarbon conversion catalysts." U.S. Patent 3,953,368. 16 pages. 1976.
4. Sinfelt, J.H. and Via, G.H. "Dispersion and structure of platinum–iridium catalysts." *J. Catal.* **56**: 1–11; 1979.
5. Kluksdahl, H.E., inventor; Chevron Research Company, assignee. "Reforming a sulfur-free naphtha with a platinum–rhenium catalyst." U.S. Patent 3,415,737. 8 pages. 1968.
6. Jacobson, R.L.; Kluksdahl, H.E.; McCoy, C.S.; and Davis, R.W. "Platinum–rhenium catalysts: a major new catalytic reforming development." *Proceedings of the American Petroleum Institute, Division of Refining.* **49**: 504–521; 1969.
7. Hansen, M. *Constitution of Binary Alloys.* 2nd ed. New York: McGraw-Hill; 1958; p. 607, 620.
8. Cullity, B.D. *Elements of X-Ray Diffraction.* Reading, MA: Addison-Wesley; 1956; p. 483, 484.
9. Prestridge, E.B.; Via, G.H.; and Sinfelt, J.H. "Electron microscopy studies of metal clusters: Ru, Os, Ru–Cu, and Os–Cu." *J. Catal.* **50**: 115–123; 1977.
10. Sinfelt, J.H.; Lam, Y.L.; Cusumano, J.A.; and Barnett, A.E. "Nature of ruthenium–copper catalysts." *J. Catal.* **42**: 227–237; 1976.
11. Sinfelt, J.H. "Heterogeneous catalysis: some recent developments." *Science.* **195**: 641–646; 1977.
12. Sinfelt, J.H. "Structure of metal catalysts." *Rev. Mod. Phys.* **51**: 569–589; 1979.
13. Kincaid, B.M. and Eisenberger, P. "Synchrotron radiation studies of the *K*-edge photoabsorption spectra of Kr, Br_2 and $GeCl_4$: a comparison of theory and experiment." *Phys. Rev. Lett.* **34**: 1361–1364; 1975.
14. Sayers, D.E.; Lytle, F.W.; and Stern, E. "New technique for investigating noncrystalline structures: Fourier analysis of the extended X-ray absorption fine structure." *Phys. Rev. Lett.* **27**: 1204–1207; 1971.
15. Kronig, R.deL. "Zur theorie der feinstruktur in den Röntgenabsorptionspektren." *Z. Phys.* **70**: 317; 1931.
16. Kronig, R.deL. "Zur theorie der feinstruktur in den Röntgenabsorptionspektren." *Z. Phys.* **75**: 191–210; 1932.
17. Kronig, R.deL. "Zur theorie der feinstruktur in den Röntgenabsorptionspektren." *Z. Phys.* **75**: 468–475; 1932.
18. Lytle, F.W.; Sayers, D.; and Stern, E. "Extended X-ray absorption fine structure technique. II. Experimental practice and selected results." *Phys. Rev.* **B11**: 4825–4835; 1975.
19. Via, G.H.; Sinfelt, J.H.; and Lytle, F.W. "Extended X-ray absorption fine structure (EXAFS) of dispersed metal catalysts." *J. Chem. Phys.* **71**: 690–699; 1979.
20. Sayers, D.E.; Lytle, F.W.; and Stern, E. "Point scattering theory of X-ray *K*-absorption fine structure." In: Henke, B.L.; Newkirk, J.B.; and Mallett, G.R.; eds.

Advances in X-Ray Analysis. Vol. 13. New York: Plenum; 1970; p. 248–271.

21. Stern, E. "Theory of the extended X-ray absorption fine structure." *Phys. Rev.* **B10**: 3027–3037; 1974.
22. Ashley, C.A. and Doniach, S. "Theory of extended X-ray absorption edge fine structure (EXAFS) in crystalline solids." *Phys. Rev.* **B11**: 1279–1288; 1975.
23. Lee, P.A. and Pendry, J.B. "Theory of the extended X-ray absorption fine structure." *Phys. Rev.* **B11**: 2795–2811; 1975.
24. Lee, P.A. and Beni, G. "New method for the calculation of atomic phase shifts: application to extended X-ray absorption fine structure (EXAFS) in molecules and crystals." *Phys. Rev.* **B15**: 2862–2883; 1977.
25. Stern, E.; Sayers, D.; and Lytle, F.W. "Extended X-ray absorption fine structure technique. III. Determination of physical parameters." *Phys. Rev.* **B11**: 4836–4846; 1975.
26. Warren, B.E. *X-Ray Diffraction.* Reading, MA: Addison-Wesley; 1969; p. 217.
27. Sayers, D.E. "A new technique to determine amorphous structure using extended X-ray absorption fine structure." Seattle: University of Washington; 1971. Ph.D. dissertation.
28. Lytle, F.W.; Via, G.H.; and Sinfelt, J.H. "X-ray absorption spectroscopy: catalyst applications." In: Winick, H. and Doniach, S., eds. *Synchrotron Radiation Research.* New York: Plenum; 1980; p. 401–424.
29. Fraser, R.D.B. and Suzuki, E. "Resolution of overlapping absorption bands by least squares procedures." *Anal. Chem.* **38**: 1770–1773; 1966.
30. Stone, H. "Mathematical resolution of overlapping spectral lines." *J. Opt. Soc. Am.* **52**: 998–1003; 1962.
31. Sinfelt, J.H.; Via, G.H.; and Lytle, F.W. "Structure of bimetallic clusters. Extended X-ray absorption fine structure (EXAFS) studies of Ru–Cu clusters." *J. Chem. Phys.* **72**: 4832–4844; 1980.
32. Sinfelt, J.H.; Via, G.H.; Lytle, F.W.; and Greegor, R.B. "Structure of bimetallic clusters. Extended X-ray absorption fine structure (EXAFS) studies of Os–Cu clusters." *J. Chem. Phys.* **75**: 5527–5537; 1981.
33. MacGillavry, Caroline H.; Rieck, Gerard D.; and Lonsdale, Kathleen, eds. *International Tables for X-ray Crystallography.* Vol. III. Birmingham, England: Kynoch; 1962; p. 282.
34. Teo, B.K. and Lee, P.A. "Ab initio calculations of amplitude and phase functions for extended X-ray absorption fine structure spectroscopy." *J. Amer. Chem. Soc.* **101**: 2815–2832; 1979.
35. Citrin, P.H.; Eisenberger, P.; and Kincaid, B.M. "Transferability of phase shifts in extended X-ray absorption fine structure." *Phys. Rev. Lett.* **36**: 1346–1349; 1976.

36. Pauling, L. *Nature of the Chemical Bond.* 2nd ed. Ithaca, NY: Cornell University Press; 1948; p. 410.

37. Mott, N.F. "The basis of the electron theory of metals with special reference to the transition metals." *Proc. Phys. Soc. (London).* **62**: 416–422; 1949.

38. Cauchois, Y. and Mott, N.F. "The interpretation of X-ray absorption spectra of solids." *Philos. Mag.* **40**: 1260–1269; 1949.

39. Lytle, F.W.; Wei, P.S.P.; Greegor, R.B.; Via, G.H.; and Sinfelt, J.H. "Effect of chemical environment on magnitude of X-ray absorption resonance at L_{III} edges. Studies on metallic elements, compounds, and catalysts." *J. Chem. Phys.* **70**: 4849–4855; 1979.

40. Carter, J.L.; McVicker, G.B.; Weissman, W.; Kmak, W.S.; and Sinfelt, J.H. "Bimetallic catalysts; application in catalytic reforming." *Applied Catal.* **3**: 327–346; 1982.

41. Garten, R.L. and Sinfelt, J.H. "Structure of Pt–Ir catalysts: Mössbauer spectroscopy studies employing ^{57}Fe as a probe." *J. Catal.* **62**: 127–139; 1980.

42. Sinfelt, J.H. "Heterogeneous catalysis by metals." *Prog. Solid State Chem.* **10** *(2)*: 55–69; 1975.

43. McVicker, G.B.; Baker, R.T.K.; Garten, R.L.; and Kugler, E.L. "Chemisorption properties of iridium on alumina catalysts." *J. Catal.* **65**: 207–220; 1980.

44. Sinfelt, J.H. "Specificity in catalytic hydrogenolysis by metals." *Advan. Catal.* **23**: 91–119; 1973.

45. Pearson, W.B. *A Handbook of Lattice Spacings and Structures of Metals and Alloys.* New York: Pergamon Press; 1964; p. 704.

46. Klug, H.P. and Alexander, L.E. *X-ray Diffraction Procedures for Polycrystalline and Amorphous Materials.* 2nd ed. New York: Wiley; 1954; p. 656.

47. Raub, E. and Platte, W. "Tempering and decomposition of platinum–iridium alloys." *Z. Metallk.* **47**: 688–693; 1956.

48. Sinfelt, J.H.; Via, G.H.; and Lytle, F.W. '"Structure of bimetallic clusters. Extended X-ray absorption fine structure (EXAFS) studies of Pt–Ir clusters." *J. Chem. Phys.* **76**: 2779–2789; 1982.

49. Bond, G.C. *Catalysis by Metals.* London and New York: Academic Press; 1962; p. 489, 490.

50. Swalin, R.A. *Thermodynamics of Solids.* New York: Wiley; 1962; p. 193.

51. Kuijers, F.J. and Ponec, V. "The surface composition of platinum–iridium alloys." *Appl. Surface Sci.* **2**: 43–54; 1978.

52. Sinfelt, J.H. "Catalytic reforming of hydrocarbons." In: Anderson, John R. and Boudart, Michel, eds. *Catalysis–Science and Technology.* Vol. 1. Berlin, Heidelberg: Springer-Verlag; 1981; p. 257–300.

53. Frauenfelder, H. *The Mössbauer Effect.* New York: Benjamin; 1962; p. 1–96.
54. Garten, R.L. "Direct evidence for bimetallic clusters." *J. Catal.* **43**: 18–33; 1976.
55. Vannice, M.A. and Garten, R.L. "The synthesis of hydrocarbons from CO and H_2 over well–characterized supported PtFe catalysts." *J. Molec. Catal.* **1**: 201–222; 1975/76.
56. Garten, R.L. and Ollis, D.F. "The chemical state of iron in reduced PdFe/Al_2O_3 catalysts." *J. Catal.* **35**: 232–246; 1974.
57. Bartholomew, C.H. and Boudart, M. "Surface composition and chemistry of supported platinum–iron alloys." *J. Catal.* **29**: 278–291; 1973.
58. Stevens, J.G. and Stevens, V.E. *Mössbauer Effect Data Index.* New York: Plenum; 1973; p. 156.
59. Vannice, M.A.; Lam, Y.L.; and Garten, R.L. "CO hydrogenation over well-characterized Ru–Fe alloys." *Preprints, American Chemical Society Division of Petroleum Chemistry.* **23**: 495–501; 1978.
60. Wertheim, G.K. "Mössbauer effect: applications to magnetism." *J. Appl. Phys. (Supp.).* **32**: 110S–117S; 1961.
61. Mizoguchi, T.; Sasaki, T.; and Chikazumi, S. "Appearance of ferromagnetism in Fe–Ir–Pt alloys." *American Institute of Physics Conference Proceedings.* **5:** 445–460; 1972.
62. Kanashiro, M. and Kunitomi, N. "Mössbauer effect of ^{193}Ir and ^{57}Fe in Fe–Pt–Ir alloys." *J. Phys. Soc. Japan.* **43**: 1559–1568; 1977.
63. Johnson, M.F.L. and Leroy, V.M. "The state of rhenium in Pt/Re/Alumina catalysts." *J. Catal.* **35**: 434–440; 1974.
64. Webb, A.N. "Reducibility of supported rhenium." *J. Catal.* **39**: 485–486; 1975.
65. Johnson, M.F.L. "The state of rhenium in Pt/Re/Alumina catalyst." *J. Catal.* **39:** 487; 1975.
66. McNicol, B.D. "The reducibility of rhenium in Re on γ-alumina and Pt–Re on γ-alumina catalysts." *J. Catal.* **46**: 438–440; 1977.
67. Bolivar, C.; Charcosset, H.; Frety, R.; Primet, M.; Tournayan, L.; Betizeau, C.; Leclercq, G.; and Maurel, R. "Platinum–rhenium/alumina catalysts. I. Investigation of reduction by hydrogen." *J. Catal.* **39**: 249–259; 1975.
68. Bolivar, C.; Charcosset, H.; Frety, R.; Primet, M.; Tournayan, L.; Betizeau, C.; Leclercq, G.; and Maurel, R. "Platinum–rhenium–alumina catalysts. II. Study of the metallic phase after reduction." *J. Catal.* **45**: 163–178; 1976.
69. Betizeau, C.; Leclercq, G.; Maurel, R.; Bolivar, C.; Charcosset, H.; Frety, R.; and Tournayan, L. "Platinum–rhenium–alumina catalysts. III. Catalytic properties." *J. Catal.* **45**: 179–188; 1976.
70. Tournayan, L.; Charcosset, H.; Frety, R.; Leclercq, C.; Turlier, P.; Barbier, J.; and Leclercq, G. "Hydrogen–oxygen titrations over platinum–rhenium and plati-

num–iridium/alumina catalysts in a vacuum microbalance." *Thermochimica Acta.* **27**: 95–110; 1978.

71. Charcosset, H.; Frety, R.; Leclercq, G.; Mendes, E.; Primet, M.; and Tournayan, L. "The state of Re in Pt–Re/γ-Al_2O_3 catalysts." *J. Catal.* **56**: 468–471; 1979.
72. Charcosset, H. "The state of the metal phase in supported platinum-containing bimetallic catalysts." *Platinum Metals Rev.* **23***(1)*: 18–21; 1979.
73. Biloen, P.; Helle, J.N.; Verbeek, H.; Dautzenberg, F.M.; and Sachtler, W.M.H. "The role of rhenium and sulfur in platinum-based hydrocarbon-conversion catalysts." *J. Catal.* **63**: 112–118; 1980.

ABSTRACT

Bimetallic catalysts have had a major industrial impact, specifically for the reforming of petroleum naphtha fractions to produce high octane number components for gasolines. The reforming process is conducted at high temperatures (425–525°C) and high pressures (10–35 atm). The objective of reforming is the selective conversion of saturated hydrocarbons to aromatic hydrocarbons, which are excellent antiknock components for gasolines and are also of interest as petrochemicals. During the 1950s and 1960s, platinum catalysts dominated catalytic reforming. In the 1970s, however, bimetallic catalysts widely replaced the traditional platinum catalysts. The advantages of the new bimetallic catalysts include higher activity, much improved activity maintenance, and higher reformate yields. Two bimetallic catalysts that have been applied extensively in commercial reformers are platinum–rhenium on alumina and platinum–iridium on alumina. In some respects, these two catalysts perform similarly relative to the earlier platinum on alumina catalysts. However, there are differences in detail in the nature of the catalysts and in their reforming properties. The differences are such that the two catalysts can be combined effectively in a reforming operation to take advantage of the desirable properties of each.

Chapter 5

INDUSTRIAL APPLICATIONS OF BIMETALLIC CATALYSTS

Bimetallic catalysts have had a major impact in industrial catalysis, notably in the catalytic reforming of petroleum fractions. As background for a discussion of the application of such catalysts in the reforming process, some of the major features of catalytic reforming are reviewed first.

5.1 DESCRIPTION OF CATALYTIC REFORMING

Over the past three decades catalytic reforming has evolved very rapidly. It is now one of the most important industrial applications of catalysis. The reforming process was originally developed to produce gasoline components of high antiknock quality to meet the fuel requirements of high compression ratio automobile engines.

The objective of the process is to convert saturated hydrocarbons (alkanes and cycloalkanes) in petroleum naphtha fractions to aromatic hydrocarbons as selectively as possible, since the latter have excellent antiknock ratings *(1,2)*. Naphtha fractions are composed of hydrocarbons with boiling points in the approximate range of 50–200°C. Reaction temperatures of 425–525°C and pressures of 10–35 atm are employed in the process. Reforming catalysts commonly contain platinum *(3–5)* or a combination of platinum and a second metallic element such as rhenium *(6)* or iridium *(2,7)*.

A typical catalytic reforming unit consists of a number of fixed-bed reactors, frequently four, in series. The naphtha feedstock is vaporized and heated to the desired reaction temperature, then admitted to the first reactor. As the components in the naphtha undergo reaction during passage through the catalyst bed, the temperature of the vapor stream decreases by 70–100°C due to the endothermicity of the reaction. The major reaction occurring in the first catalyst bed is the dehydrogenation of cycloalkanes to aromatics.

The effluent from the first reactor is reheated to the desired reaction temperature and admitted to the second reactor. As reaction occurs, the temperature of the vapor stream decreases again, although to a smaller extent than in the first reactor. The process of reheating followed by reaction is repeated until the hydrocarbon stream has passed through all of the reactors. The effluent from the final reactor is cooled and separated into liquid and gaseous products.

The liquid product, known as reformate, consists essentially of C_5 through about C_{10} hydrocarbons. Aromatic hydrocarbons in the liquid product are typically present in amounts in the range of 60–70 wt%, the actual amount depending on the reaction conditions. The antiknock quality of the liquid reformate increases with increasing aromatics content.

The gaseous product consists of hydrogen and C_1–C_4 hydrocarbons, with a hydrogen concentration that is commonly within the range of 60–90 mole%. A stream of this gas is recycled to the inlet of the first reactor, where it is combined with the naphtha feed in a ratio of about 5–10 moles of recycle gas per mole of naphtha. Consequently, there is a high partial pressure of hydrogen in the system. The high hydrogen partial pressure is crucial for the maintenance of catalytic activity, since it retards fouling of the catalyst surface by hydrocarbon residues.

Such residues are formed via extensive dehydrogenation of chemisorbed hydrocarbons to highly unsaturated species which, in turn, readily undergo condensation or polymerization reactions. The hydrogen inhibits the formation of highly unsaturated species on the surface and also removes hydrocarbon residues via hydrogenolysis reactions *(8–10)*. Since increasing hydrogen pressure also has the effect of decreasing the yield of aromatic hydrocarbons, the choice of hydrogen pressure for a reformer is a balance of product yields against catalyst deactivation rate.

While catalytic reforming is a complex process involving reactions of a large number of hydrocarbons, a reasonable understanding of the chemistry of the process and of the functioning of reforming catalysts has evolved. This understanding has resulted from studies of the reactions of individual hydrocarbons and from investigations of reforming catalysts by a variety of chemical and physical methods.

5.1.1 Major Reactions in Catalytic Reforming

The major reactions in catalytic reforming are (a) dehydrogenation of cyclohexanes to aromatic hydrocarbons, (b) dehydroisomerization of alkylcyclopentanes to aromatic hydrocarbons, (c) dehydrocyclization of alkanes to aromatic hydrocarbons, (d) isomerization of *n*-alkanes to branched alkanes, and (e) fragmentation reactions (hydrocracking and hydrogenolysis) of alkanes and cycloalkanes to low molecular weight alkanes *(1,11).* The reactions are illustrated in Figure 5.1 with specific examples.

The catalysts on which these reactions occur are bifunctional; that is, they possess two different types of catalytic activity. In addition to catalyzing hydrogenation and dehydrogenation reactions, they also catalyze hydrocarbon rearrangements of a type commonly observed in acid catalysis. These two distinct types of activity are associated with two different components of the catalyst, at least under conditions typical of catalytic reforming.

The hydrogenation and dehydrogenation activities are associated with a metal component (e.g., platinum), while the activity for acid catalysis is associated with the presence of acidic sites on the surface of the carrier on which the metal component is dispersed. The carrier is commonly alumina, which is covered by the metal component to an extent of about 1% or less in the catalysts typically employed in reforming.

Early studies of reactions (*a*) through (*e*) on bifunctional catalysts were published by Haensel and Donaldson *(5),* Ciapetta and Hunter *(12,13),* Heinemann et al. *(14),* and Hettinger et al. *(15).*

The major part of the antiknock improvement obtained in the catalytic

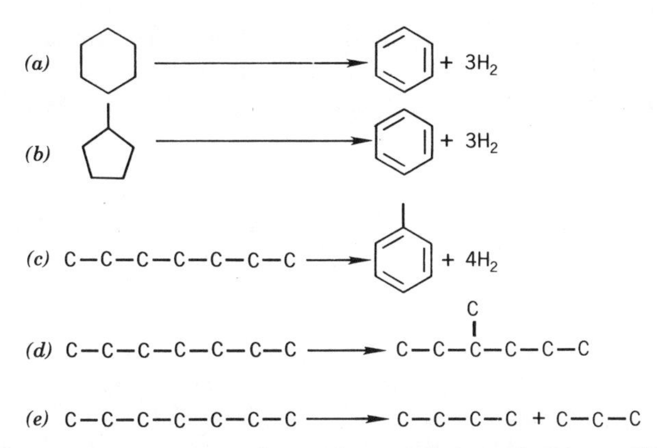

Figure 5.1 Major reactions in catalytic reforming illustrated with specific examples: (*a*) dehydrogenation of cyclohexanes to aromatic hydrocarbons; (*b*) dehydroisomerization of alkylcyclopentanes to aromatic hydrocarbons; (*c*) dehydrocyclization of alkanes to aromatic hydrocarbons; (*d*) isomerization of *n*-alkanes to branched alkanes; (*e*) fragmentation reactions (hydrocracking and hydrogenolysis) yielding low carbon number alkanes.

reforming of petroleum naphthas is due to the formation of aromatic hydrocarbons. Lesser contributions result from the isomerization of *n*-alkanes to branched alkanes and from hydrocracking of hydrocarbons to lower molecular weight hydrocarbons in the gasoline boiling range.

Of the major reactions occurring in catalytic reforming, the dehydrogenation of cyclohexanes to aromatic hydrocarbons occurs most readily. Isomerization reactions also occur readily, but not nearly so fast as the dehydrogenation of cyclohexanes. The limiting reactions in most catalytic reforming operations are dehydrocyclization and hydrocracking, which generally occur at much lower rates.

The thermodynamics of the more important reactions in catalytic reforming can be discussed conveniently by referring to the equilibria involved in various interconversions among C_6 hydrocarbons. Some thermodynamic equilibrium constants at 500°C, a typical temperature in catalytic reforming, and heats of reaction are given in Table 5.1 *(1)*. The equilibrium constant K in Table 5.1 is defined in terms of the partial pressures of the reactants and products expressed in atmospheres.

The dehydrogenation of cyclohexane and the dehydrocyclization of *n*-hexane to yield benzene are strongly endothermic, so that increasing temperature markedly improves the extent of conversion to benzene. Hydrogen partial pressure obviously has a marked effect on the extent of formation of benzene, and from the viewpoint of equilibria alone, it is advantageous to operate at as high a temperature and as low a hydrogen partial pressure as possible to maximize the yield of the aromatic hydrocarbon.

However, other considerations, such as catalyst deactivation due to for-

Table 5.1 Thermodynamic Data on Reactions of C_6 Hydrocarbons *(1)*

Reaction	K^a	ΔH_R, kcal/mole
Cyclohexane → Benzene + 3 H_2	6×10^5	52.8
Methylcyclopentane → Cyclohexane	0.086	−3.8
n-Hexane → Benzene + $4H_2$	0.78×10^5	63.6
n-Hexane → 2-Methylpentane	1.1	−1.4
n-Hexane → 3-Methylpentane	0.76	−1.1
n-Hexane → 1-Hexene + H_2	0.037	31.0

NOTE: Reprinted with permission from Academic Press, Inc.

[a]Equilibrium constant at 500°C.

mation of carbonaceous residues on the surface, place a practical upper limit on temperature and a lower limit on hydrogen partial pressure in catalytic reforming operations. The thermodynamic considerations in the formation of aromatic hydrocarbons from higher molecular weight alkanes and cycloalkanes are qualitatively the same as for the formation of benzene from *n*-hexane and cyclohexane. Quantitatively, the equilibrium favors the aromatic hydrocarbon to a greater extent as the molecular weight increases.

The dehydrogenation of alkanes (paraffins) to alkenes (olefins) does not occur to a large extent at typical reforming conditions. For example, equilibrium conversion of *n*-hexane to 1-hexene is about 0.3% at 510°C and 17 atm hydrogen partial pressure. Nevertheless, alkane dehydrogenation is of considerable importance, since olefins appear to be intermediates in some of the reactions. This matter is discussed in more detail in a later section.

The formation of olefins from alkanes, similar to the formation of aromatic hydrocarbons, is favored by the combination of high temperature and low hydrogen partial pressure. The extent of olefin formation is important in determining the rates of those reactions that proceed via olefin intermediates. Thermodynamics sets an upper limit on the attainable concentrations of such intermediates in the system.

The equilibria for isomerization reactions are much less temperature sensitive than those for dehydrogenation reactions, since the heats of reaction are relatively small. The equilibrium between methylcyclopentane and cyclohexane favors the former, indicating that the five-membered ring structure is more stable than the six-membered ring. In the equilibria between *n*-hexane and the methylpentanes, 2-methylpentane is the favored isomer over 3-methylpentane. This is reasonable from the simple statistical consideration that the substituent methyl group can occupy either of two equivalent positions in the former molecule, compared to one in the latter.

Although thermodynamic calculations show appreciable quantities of dimethylbutanes at equilibrium (about 30–35% of the total hexane isomers at 500°C), such quantities are not observed in reforming. Equilibria are established readily between *n*-hexane and the methylpentanes, but not between these hydrocarbons and the dimethylbutanes. The reaction kinetics are not favorable for the rearrangement of singly branched to doubly branched isomers *(11)*. This limitation apparently does not exist for the rearrangement of the normal structure to the singly branched structures.

Hydrocracking and hydrogenolysis reactions involve rupture of carbon–carbon bonds followed by hydrogenation. Such reactions are very exothermic and are highly favorable from a thermodynamic point of view. Thermodynamic considerations are not important in determining the extent to which these reactions occur in catalytic reforming. Practically speaking, the reactions are limited solely by kinetic factors.

5.1.2 Mechanistic Considerations

In discussing the mechanism of hydrocarbon transformations on bifunctional catalysts, it is useful to refer to a reaction scheme originally proposed by Mills et al. *(16)* to describe the reforming of C_6 hydrocarbons. The scheme is shown in Figure 5.2. The vertical reaction paths in the figure take place on the hydrogenation–dehydrogenation centers of the catalyst and the horizontal reaction paths on the acidic centers.

According to the mechanism, the conversion of methylcyclopentane to benzene, for example, first involves dehydrogenation to methylcyclopentenes on hydrogenation–dehydrogenation centers of the catalyst, followed by isomerization to cyclohexene on acidic centers. The cyclohexene then returns to hydrogenation–dehydrogenation centers where it can either be hydrogenated to cyclohexane or further dehydrogenated to benzene, the relative amounts of these products depending on reaction conditions.

The mode of transport of olefin intermediates between metal and acidic sites must be considered in this type of reaction scheme. In the isomerization of alkanes or dehydroisomerization of alkylcyclopentanes, a reaction sequence involving transport of olefin intermediates from metal to acidic

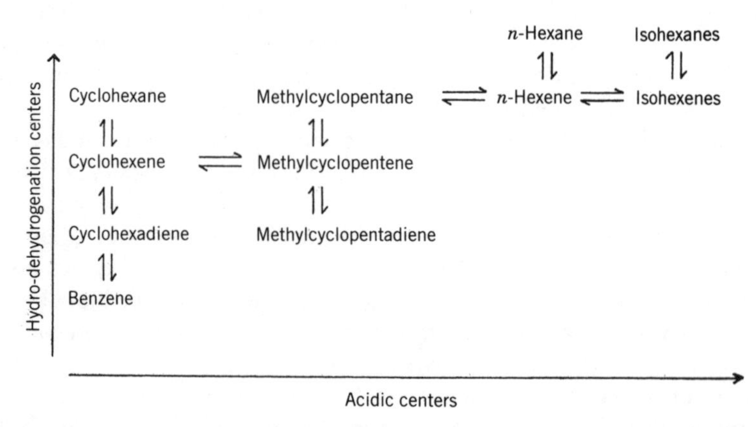

Figure 5.2 Reaction paths in catalytic reforming of C_6 hydrocarbons. (Reprinted with permission from Ref. 16 [Mills et al. (1953)]. Copyright 1953 American Chemical Society.)

sites via the gas phase gives a good account of much experimental kinetic data *(17–19).*

Investigations of the isomerization of alkanes in recent years have provided evidence that the reaction can occur on certain metals, notably platinum, in the absence of a separate acidic component in the catalyst *(20–22).* While it has been shown that a purely metal-catalyzed isomerization process can occur, the findings do not challenge the commonly accepted mode of action of bifunctional reforming catalysts in which separate metal and acidic sites participate in the reaction. The available data at conditions commonly employed with commercial reforming catalysts indicate that a purely metal-catalyzed process does not contribute appreciably to the overall isomerization reaction on a bifunctional catalyst.

Nevertheless, the metal-catalyzed isomerization reaction is of interest from the point of view of understanding the nature of hydrocarbon transformations on metal surfaces. It has been suggested that carbonium-ion-like intermediates are involved in alkane isomerization reactions on platinum *(23),* and a specific mechanism has been proposed by Anderson and Avery *(24).*

In the reaction scheme in Figure 5.2, the dehydrocyclization of *n*-hexane proceeds via formation of *n*-hexene on dehydrogenation centers, followed by cyclization of the *n*-hexene to methylcyclopentane on acidic centers. The methylcyclopentane then is converted to benzene in the manner already described. Alternatively, it seems possible that a hexadiene may be an intermediate in the reaction sequence. Such a sequence would involve formation of a hexadiene on platinum sites, followed by cyclization on acidic centers to form a cyclic olefin, methylcyclopentene *(1).*

However, there is also evidence that dehydrocyclization may proceed by another route involving only the metal component of the catalyst. It has been observed that unsupported platinum powders catalyze the dehydrocyclization of *n*-heptane *(21).* Also, Dautzenberg and Platteeuw *(25)* report that dehydrocyclization of *n*-hexane to benzene occurs over a catalyst in which platinum is supported on a nonacidic alumina. Since bifunctional catalysis with participation of acidic sites is then presumably eliminated, the activity is attributed to the platinum itself.

Dautzenberg and Platteeuw also suggest that dehydrocyclization may occur via the gas phase cyclization of a hexatriene formed on the platinum *(25).* However, the rate of dehydrocyclization, which in general depends on hydrogen partial pressure, is a particularly strong inverse function of this variable for the path involving a hexatriene intermediate. Consequently, this path is probably not an important contributor to the overall dehydrocyclization reaction at the high hydrogen partial pressures typical of catalytic reforming.

Reactions involving rupture of carbon–carbon bonds with accompanying hydrogenation (i.e., hydrocracking and hydrogenolysis) occur to a significant extent in reforming. When such reactions involve only the metal component of the catalyst, we arbitrarily use the term hydrogenolysis in referring to them. Hydrogenolysis reactions have been studied extensively in recent years *(26).* The sequence of steps generally postulated for the reaction includes (a) chemisorption of a hydrocarbon reactant with dissociation of carbon–hydrogen bonds to form an unsaturated hydrocarbon residue, and (b) rupture of carbon–carbon bonds in this surface residue. The species formed in step (b) are then hydrogenated to form the products of the reaction.

When the rupture of carbon–carbon bonds involves acidic sites on the carrier, we arbitrarily designate the reaction hydrocracking. Saturated hydrocarbons presumably undergo hydrocracking by two different routes, one involving only the acidic sites of the carrier *(27,28),* the other involving a cooperative action *(29)* of these sites with metal sites. In the latter route, olefins are regarded as intermediates in a reaction sequence similar to that already described for isomerization, except that the olefin undergoes a fragmentation (cracking) reaction rather than a skeletal rearrangement.

5.2 REFORMING WITH BIMETALLIC CATALYSTS

During the 1950s and 1960s platinum–alumina catalysts dominated catalytic reforming. In the 1970s, however, bimetallic catalyst systems were introduced widely *(6,7,30–32).* The advantages of the new catalysts include higher activity, much improved activity maintenance, and higher reformate yields. Two catalyst systems that have been applied extensively in commercial reformers are platinum–rhenium on alumina and platinum–iridium on alumina.

The higher activity and better activity maintenance of a platinum–iridium on alumina catalyst relative to a platinum on alumina catalyst are illustrated in Figure 5.3. Data are shown for the reforming of a 50–200°C boiling range Venezuelan naphtha at 487°C and 14.6 atm over a catalyst containing 0.3 wt% each of platinum and iridium and a catalyst containing 0.6 wt% platinum. The naphtha weight hourly space velocity (grams of naphtha per hour per gram of catalyst) was 2.1, and the mole ratio of hydrogen to naphtha at the reactor inlet was approximately 6. The naphtha contained (on a volume percentage basis) 47.3% alkanes, 42.2% cycloalkanes, and 10.5% aromatic hydrocarbons, and it had a density of 0.7605 g/cm^3 *(33).*

The data show the research octane number of the C_5+ liquid reformate

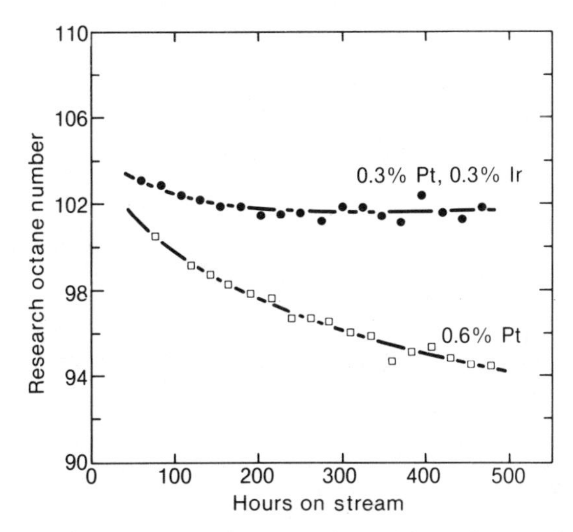

Figure 5.3 Comparison of alumina-supported platinum–iridium and platinum catalysts showing the research octane number of the liquid reformate as a function of time on stream in the reforming of a 50–200°C boiling range Venezuelan naphtha at 487°C and 14.6 atm *(33)*. (Reprinted with permission from Elsevier Scientific Publishing Company.)

as a function of time on stream. The octane number, which is largely determined by the aromatic hydrocarbon content of the reformate, reflects the activity of the catalyst. As shown in Figure 5.3, the octane number is substantially higher for the platinum–iridium catalyst than for the platinum catalyst. Also, the decline of the octane number with time on stream is lower for the platinum–iridium catalyst, an indication of its better activity maintenance. The lower rate of deactivation is also a key feature of platinum–rhenium catalysts.

In commercial practice a reformer is operated to produce a constant octane number product. As the catalyst deactivates, the temperature of the system may be increased to compensate for the lower activity. In this way the octane number of the product can be maintained at the desired level. Illustrative data for alumina-supported platinum, platinum–rhenium, and platinum–iridium catalysts in this type of operation are given in Figure 5.4 *(7)*.

In Figure 5.4, the temperature required to produce 100 research clear octane number reformate is shown as a function of time on stream in the reforming of a 99–171°C boiling range naphtha containing approximately 43, 45, and 12% by volume, respectively, of cycloalkanes, alkanes, and aromatic

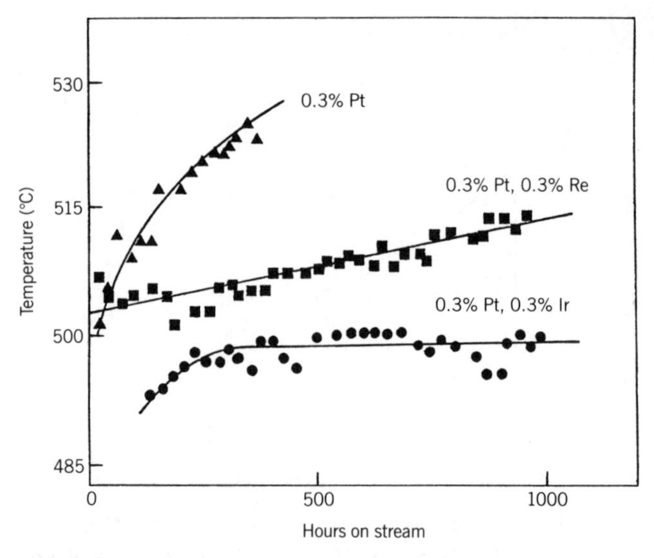

Figure 5.4 Data on the reforming of a 99–171°C boiling range naphtha showing the temperature required to produce 100 octane number product as a function of time on stream for alumina-supported platinum, platinum–rhenium, and platinum–iridium catalysts at 14.6 atm pressure *(2,7)*.

hydrocarbons. The reactor pressure was 14.6 atm, and the naphtha weight hourly space velocity was 3. Hydrogen was introduced with the naphtha at a rate of 5000 standard cubic feet per barrel of naphtha, so that the molar ratio of hydrogen to hydrocarbon at the reactor inlet was approximately 6. The reactor was surrounded by a fluidized solids bath for temperature control. The temperatures in the figure are bath temperatures.

As shown in Figure 5.4, the temperature required to maintain 100 research clear octane number product varies with the catalyst and is also a function of time on stream (the higher the catalytic activity, the lower the temperature required to produce the desired octane number reformate).

The data presented in Figures 5.3 and 5.4 are concerned with catalyst activity and the maintenance of activity during reforming. While these aspects of the performance of a reforming catalyst are important, they are not the only considerations of interest. The yield of C_5+ reformate is also very important and frequently is the crucial consideration in the choice of a catalyst. Yields of individual aromatic hydrocarbons such as benzene, toluene, and paraxylene are also important.

Consequently, it is of interest to consider the reforming properties of platinum–rhenium and platinum–iridium catalysts in more detail. First, we consider some information obtained from studies on the reforming of pure hydrocarbons over these catalysts. This information provides us with a better understanding of the way these catalysts function in reforming. After the discussion of the studies on reforming of pure hydrocarbons, the results of some extended naphtha reforming runs on these catalysts are considered in detail.

5.2.1 Reforming of Pure Hydrocarbons *(33)*

Data on rates of dehydrocyclization r_D and cracking r_C of n-heptane at 495°C and 14.6 atm are given in Table 5.2 for platinum–iridium on alumina and platinum–rhenium on alumina catalysts, and also for catalysts containing platinum or iridium alone on alumina *(33)*. The rate r_D refers to the rate of production of toluene and C_7 cycloalkanes, the latter consisting primarily of methylcyclohexane and dimethylcyclopentanes. The rate of cracking is the rate of conversion of n-heptane to C_6 and lower carbon number alkanes.

Table 5.2 n-Heptane Conversion on Pt–Ir, Pt–Re, and Related Catalysts[a] *(33)*

	Rate (mole/hr/g catalyst)[c]		Selectivity
Catalyst[b]	r_D	r_C	r_D/r_C
0.3% Ir	0.051	0.103	0.50
0.3% Pt, 0.3% Ir	0.051	0.092	0.55
0.3% Pt	0.021	0.050	0.42
0.3% Pt, 0.3% Re	0.031	0.070	0.44

NOTE: Reprinted with permission from Elsevier Scientific Publishing Company.

[a]The data were obtained with n-heptane containing 0.5 ppm sulfur after 40 hours on stream at 495°C and 14.6 atm using a 5/1 mole ratio of hydrogen to n-heptane. The catalysts contained 0.9 wt% Cl as charged.

[b]Alumina was employed as the carrier in all of the catalysts.

[c]The quantity r_D refers to the rate of dehydrocyclization of n-heptane to toluene and C_7 cycloalkanes, while r_C is the rate of cracking to C_6 and lower carbon number alkanes. The rates were determined at conversion levels of 7–12%.

The reaction rates were determined at low conversion levels (7–12%) in an attempt to minimize the effects of secondary reactions. The inlet stream to the reactor contained five moles of hydrogen per mole of *n*-heptane. The *n*-heptane contained 0.5 ppm sulfur, and the reaction rates were determined after 40 hours on stream. The catalysts contained 0.9 wt% chlorine as charged. Prior to the runs the catalysts were contacted with an H_2S-containing gas until H_2S was detected at the reactor outlet *(34).* This step is routinely employed with platinum–rhenium and platinum–iridium catalysts to suppress hydrogenolysis activity *(33).*

The rates of dehydrocyclization of *n*-heptane for the iridium and platinum–iridium catalysts are more than twice as high as the rate for the platinum catalyst, and almost twice as high as the rate for the platinum–rhenium catalyst. The rates of cracking are also higher for the iridium and the platinum–iridium catalysts.

The selectivity, defined as the ratio r_D/r_C of the rate of dehydrocyclization to the rate of cracking, is higher for the iridium-containing catalysts. It is significant that the selectivity of the platinum–iridium catalyst is higher than that of the catalyst containing iridium alone. The selectivity of the platinum–rhenium catalyst appears to be slightly higher than that of the catalyst containing platinum alone.

In Table 5.3 data are presented for the conversion of methylcyclopentane at 500°C and 14.6 atm over the same catalysts for which data on *n*-heptane reactions are presented in Table 5.2, and also for a rhenium on alumina catalyst *(33).* The selectivity to benzene and to C_1–C_6 alkanes, expressed as percentages of the total conversion of methylcyclopentane to all products, is shown after 0.5 hour and after 24 hours on stream. The methylcyclopentane weight hourly space velocity was 40 grams per hour per gram of catalyst. The inlet stream to the reactor contained five moles of hydrogen per mole of methylcyclopentane. The methylcyclopentane contained 1 ppm sulfur, and the catalysts were pretreated in the same manner as they were in the *n*-heptane conversion studies.

The selectivity to benzene is much lower for the iridium catalyst than for any of the other catalysts except rhenium on alumina. The platinum–iridium catalyst is clearly more selective than the iridium catalyst with respect to benzene formation. However, it is less selective for benzene formation than the catalyst containing platinum alone, although it is possible that this debit may disappear as the catalyst ages during a run. Also, except for the initial reaction period, the platinum–iridium catalyst is less selective for benzene formation than the platinum–rhenium catalyst.

The selectivity of the platinum–rhenium catalyst for benzene formation is

Table 5.3 Methylcyclopentane Conversion over Pt–Ir, Pt–Re, and Related Catalysts[a] *(33)*

			% Selectivity to	
Catalyst[b]	Time, hr	% Conversion	Benzene	C_1–C_6 Alkanes
0.3% Ir	0.5	33	53	47
	24	42	39	61
0.3% Pt, 0.3% Ir	0.5	42	71	29
	24	43	62	38
0.3% Pt	0.5	40	79	21
	24	17	66	34
0.3% Pt, 0.3% Re	0.5	38	71	29
	24	36	76	24
0.3% Re	0.5	17	16	84
	24	13	25	75

NOTE: Reprinted with permission from Elsevier Scientific Publishing Company.

[a]The data were obtained with methylcyclopentane containing 1 ppm sulfur at 500°C and 14.6 atm using a methylcyclopentane weight hourly space velocity of 40 and a 5/1 mole ratio of hydrogen to methylcyclopentane. The catalysts contained 0.9 wt% Cl as charged.

[b]Alumina was the carrier in all of the catalysts.

initially lower than that of the platinum catalyst, but after 24 hours on stream it is higher. The maintenance of selectivity for benzene formation with time on stream is better for the platinum–rhenium catalyst than for the platinum catalyst or either of the iridium-containing catalysts. The better selectivity maintenance may be associated with a higher retention of the sulfur incorporated during the pretreatment procedure.

The studies of *n*-heptane and methylcyclopentane conversion provide insight into the advantages of platinum–iridium and platinum–rhenium catalysts over catalysts containing only one of the transition metal components, that is, platinum, iridium, or rhenium. If, for example, we consider an iridium–alumina catalyst for the reforming of a petroleum naphtha fraction, we find that it produces a substantially higher octane number reformate than a platinum on alumina catalyst under normal reforming conditions. The iridium–alumina catalyst will also exhibit a lower rate of formation of carbonaceous residues on the surface, with the result that the maintenance of activity with time will be much superior to that of a platinum–alumina catalyst.

On the basis of activity and activity maintenance, the use of an iridium–alumina catalyst in reforming appears very reasonable *(35,36)*. However, in our experience, the yields of low molecular weight alkanes (methane and ethane) are higher with an iridium–alumina catalyst than with a platinum–alumina catalyst, resulting in lower yields of C_5+ reformate. Because of the higher value of C_5+ reformate relative to products such as methane and ethane, the iridium–alumina catalyst is not used, despite its higher activity and better activity maintenance.

The inclusion of platinum with iridium in bimetallic clusters *(7)* provides a way of moderating the formation of low molecular weight alkanes while still retaining activity and activity maintenance that are much superior to those of a platinum catalyst. Our work has repeatedly indicated the desirability of this interaction between platinum and iridium in obtaining a satisfactory product distribution.

5.2.2 Extended Naphtha Reforming Studies *(33)*

Results of some extended naphtha reforming runs illustrate further the differences between platinum–iridium on alumina and platinum–rhenium on alumina catalysts observed in the reforming of pure hydrocarbons *(33)*. The results considered here should be regarded as illustrative only, since they are limited to the reforming of a series of Persian Gulf naphtha fractions in a particular range of operating conditions of commercial interest. The catalysts used in these runs contained 0.3 wt% platinum and 0.3 wt% of either iridium or rhenium.

The studies were conducted in stainless steel tubular reactors approximately 3.8 cm in diameter and about 1200 cm^3 in volume. The reactors were immersed in electrically heated fluidized solids baths. A naphtha fraction to be reformed was vaporized and heated to reaction temperature before contacting the catalyst. The reactor effluent was separated into liquid and gaseous fractions. A portion of the hydrogen-rich gaseous fraction was recycled through the reactor to simulate commercial reforming practice. The recycle gas was combined with the vaporized naphtha fraction prior to the reactor inlet. The mole ratio of recycle gas to naphtha at the reactor inlet was approximately 7 in all of the runs to be discussed here.

As in the studies on pure hydrocarbons discussed earlier, the catalysts were contacted with an H_2S-containing gas prior to the runs, in accordance with commercial practice *(34)*. Isopropyl chloride and water were introduced with the naphthas in controlled amounts during the runs to maintain a constant level of chlorine on the catalyst. The naphthas were desulfurized to a sulfur level of 1 ppm or lower.

The initial weight hourly space velocity of naphtha fed to the reactor (grams of naphtha per hour per gram of catalyst) ranged approximately from 0.8 to 3.2 depending on the catalyst, the naphtha being reformed, and the operating conditions *(33).* Compensation for variation of catalyst activity with time on stream was made by changing the weight hourly space velocity. Thus as the octane number decreased, it was restored to the desired value by decreasing the space velocity. While this procedure would not be used commercially, it was adopted in these studies to provide a direct, quantitative measure of catalytic activity throughout a run.

The space velocity required to produce a given octane number product is a clear measure of catalytic activity (the higher the space velocity that can be used to obtain the desired octane number, the higher the catalytic activity). In devising the activity scale used in these studies, a value of 100 was assigned to the activity of a reference catalyst that produced a certain octane number product at a standard set of conditions (space velocity, temperature, hydrogen partial pressure, naphtha partial pressure).

The activity of a test catalyst, either in the fresh state or after a substantial time on stream, is obtained simply by determining the space velocity required to produce the same octane number product at the standard conditions. Empirical relations between reformate octane number and naphtha space velocity as a function of temperature, hydrogen partial pressure, and other process variables for the reference catalyst make it possible to determine the activity of a test catalyst from data at a variety of conditions. For our purposes here, the absolute values shown for the activities are not important, since we will be concerned strictly with ratios of catalytic activities.

In the reforming runs for which results are presented in Figures 5.5 to 5.10, detailed data were obtained once each day for a 3-hour reaction period. There were approximately 60 to 90 reaction periods in each run. The individual data points in Figures 5.5 to 5.10 are averages for eight consecutive reaction periods. The midpoint of the time interval spanning a given set of eight periods was taken as the time for the data point corresponding to that interval. This procedure was adopted to simplify the presentation of the large amount of data obtained in the runs.

In Figures 5.5 and 5.6, data on the platinum–iridium and platinum–rhenium catalysts are shown for the reforming of a 70–190°C boiling range Persian Gulf naphtha to produce 98 research octane number product at a pressure of 28.2 atm and a temperature of 490°C *(33).* The naphtha contained (on a liquid volume percentage basis) 69.7% alkanes, 18.5% cycloalkanes, and 11.8% aromatic hydrocarbons. The density of the naphtha was 0.7414 g/cm^3. The data in Figure 5.5 show that the platinum–iridium catalyst is almost twice as active as the platinum–rhenium catalyst.

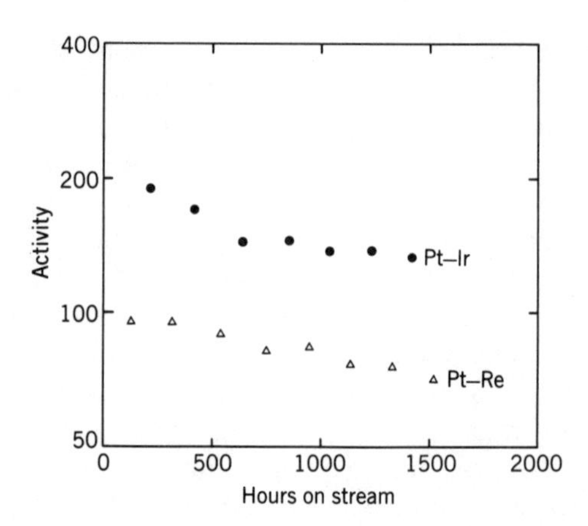

Figure 5.5 Comparison of activities of alumina-supported platinum–iridium and platinum–rhenium catalysts for the reforming of a 70–190°C boiling range Persian Gulf naphtha at 490°C and 28.2 atm to produce 98 research octane number reformate *(33).* (Reprinted with permission from Elsevier Scientific Publishing Company.)

Deposition of carbonaceous residues on the platinum–iridium catalyst was less than half of that on the platinum–rhenium catalyst (2.9 vs. 6.9 wt% on catalyst at the ends of the runs). In general, we observed that the rate of deposition of such residues on platinum–iridium catalysts is lower than on platinum or platinum–rhenium catalysts under conventional reforming conditions.

In Figure 5.6 the yields of the various products are shown. The yield of C_5+ reformate is higher for the platinum–rhenium catalyst, the average difference amounting to about 1 vol% for the final two-thirds of the runs. Correspondingly, the yields of C_1 and C_2 hydrocarbons (methane and ethane) are lower for the platinum–rhenium catalyst. The yields of C_3 and C_4 hydrocarbons are very nearly the same for the two catalysts, while the hydrogen yield is higher for the platinum–rhenium catalyst. The higher hydrogen yield is consistent with the lower yields of methane and ethane.

As a result of the higher yields of methane and ethane in the run on the platinum–iridium catalyst, the hydrogen concentration in the recycle gas stream was lower than in the run on the platinum–rhenium catalyst. Consequently, the hydrogen partial pressure at the reactor inlet was also lower. The average hydrogen partial pressures were 15.1 and 16.5 atm, respectively, for the runs on the platinum–iridium and platinum–rhenium catalysts. The difference in hydrogen partial pressure at a fixed total pressure is a consequence of the different compositions of the gaseous products, which, in turn, reflect

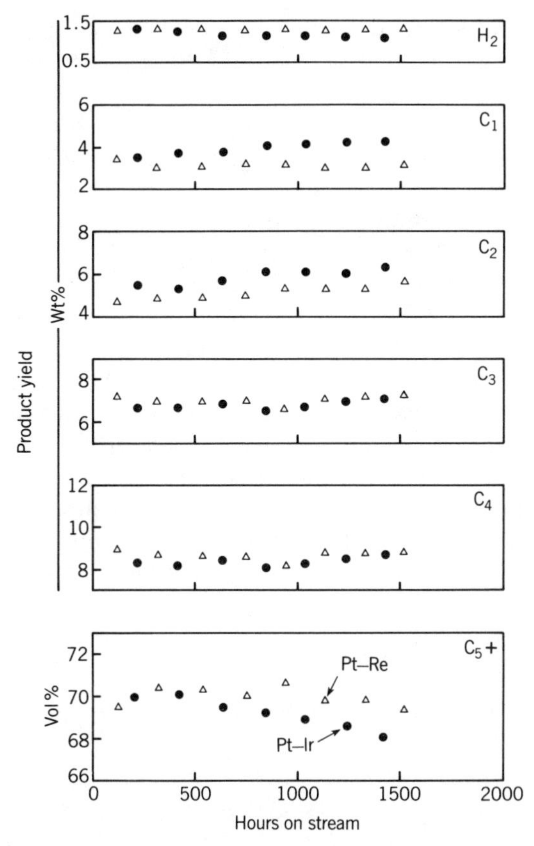

Figure 5.6 Comparison of product yields obtained with platinum–iridium and platinum–rhenium catalysts in the reforming runs for which activity data are shown in Figure 5.5 *(33)*. (Platinum–iridium catalyst, ●; platinum–rhenium catalyst, △.) (Reprinted with permission from Elsevier Scientific Publishing Company.)

differences in the kinetics of the various reactions occurring over the catalysts. This feature is characteristic of all the studies to be discussed here.

For the wide boiling range naphtha just considered, the C_5+ yield obtained with the platinum–rhenium catalyst is about 1 vol% higher than that obtained with the platinum–iridium catalyst, despite the fact that the hydrogen partial pressure is lower for the platinum–iridium catalyst. This result is typical for such wide boiling range naphthas. For light naphtha fractions such as the one to be considered next, however, the effect of the lower hydrogen partial pressure observed with the platinum–iridium catalyst can be large enough so that the C_5+ yield is higher than that obtained with the platinum–rhenium catalyst.

In Figures 5.7 and 5.8, data are presented for the reforming of a 65–150°C boiling range Persian Gulf naphtha at a temperature of 488°C to produce 96 research octane number reformate *(33).* The pressure was 28.2 atm, except for a period at 35.0 atm in the approximate time interval from hour 1460 to hour 1710 in each run. The naphtha had a density of 0.7243 g/cm^3 and contained, on a liquid volume percentage basis, 70.2% alkanes, 21.1% cycloalkanes, and 8.6% aromatic hydrocarbons.

As shown in Figure 5.7, the platinum–iridium catalyst is more than twice as active as the platinum–rhenium catalyst for most of the time on stream. In Figure 5.7 there appears to be no deactivation of the platinum–iridium catalyst out to hour 2000, and in Figure 5.3, discussed earlier, the rate of deactivation appears to be nil after hour 200. However, it should be appreciated that the platinum–iridium catalyst, like all reforming catalysts, deactivates with time. Frequently, there is an extended period of operation in which the extent of deactivation is practically nil. However, the deactivation rate in such cases

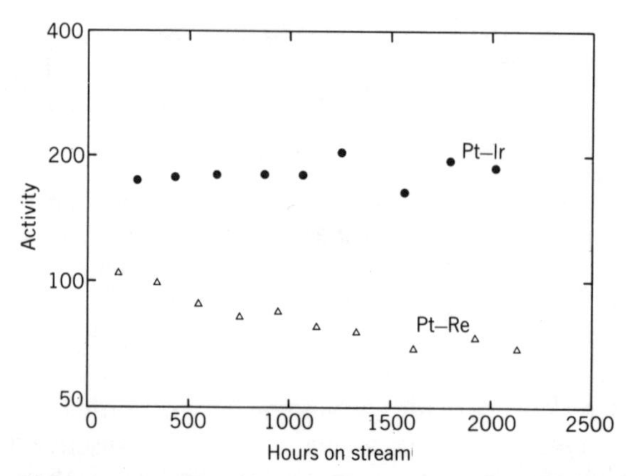

Figure 5.7 Comparison of activities of alumina-supported platinum–iridium and platinum–rhenium catalysts for the reforming of a 65–150°C boiling range Persian Gulf naphtha at 488°C and 28.2 atm to produce 96 research octane number reformate *(33).* (Data in the approximate time interval from hour 1460 to hour 1710 are for operation at 35 atm.) (Reprinted with permission from Elsevier Scientific Publishing Company.)

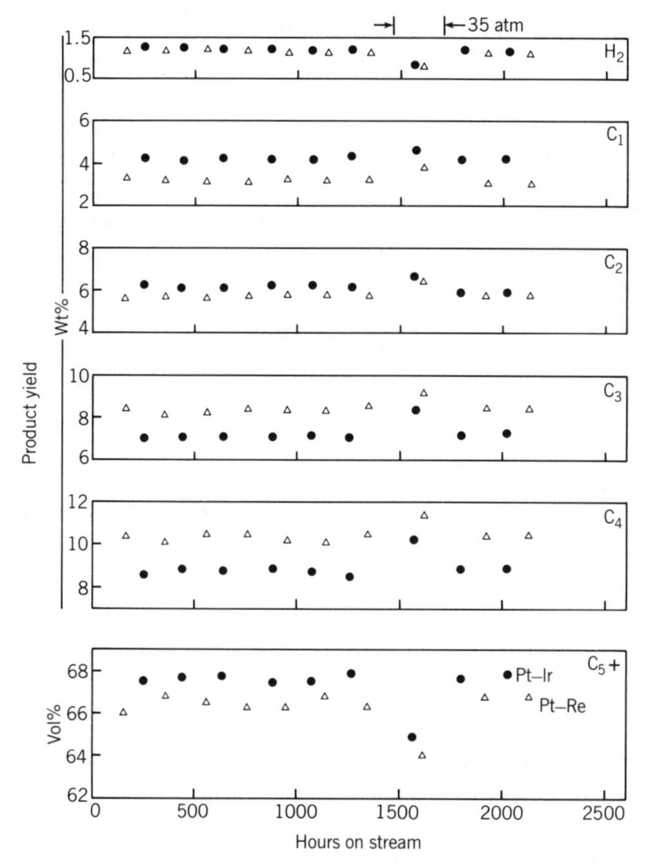

Figure 5.8 Comparison of product yields obtained with platinum–iridium and platinum–rhenium catalysts in the reforming runs for which activity data are shown in Figure 5.7 *(33).* (Platinum–iridium catalyst, ●; platinum–rhenium catalyst, △.) (Reprinted with permission from Elsevier Scientific Publishing Company.)

will eventually increase significantly with increasing time on stream, and regeneration of the catalyst will be necessary.

Data on product yields for the 65–150° Persian Gulf naphtha are shown in Figure 5.8. In contrast to the results for the 70-190°C Persian Gulf naphtha, the C_5+ yield for the platinum–iridium catalyst is at least 1 vol% higher than that for the platinum–rhenium catalyst. The yields of C_1 and C_2 hydrocarbons are still higher for the platinum–iridium catalyst than for the platinum–rhenium catalyst, but the yields of C_3 and C_4 hydrocarbons are lower. The hydrogen yields are about the same for the catalysts. The average hydrogen partial pressures were 13.5 and 15.3 atm, respectively, during the runs on the platinum–iridium and platinum–rhenium catalysts.

In general, the results of extensive investigations on the reforming properties of platinum–iridium and platinum–rhenium catalysts have shown that the activity of platinum–iridium catalysts is consistently higher than that of platinum–rhenium catalysts under conditions typical of commercial reforming operations. The activity of a 0.3 wt% Pt, 0.3 wt% Ir catalyst is generally about twice as high as that of a 0.3 wt% Pt, 0.3 wt% Re catalyst for naphtha reforming operations.

With regard to product distribution, platinum–iridium catalysts usually give slightly higher yields of methane and ethane than do platinum–rhenium catalysts. Correspondingly, the yield of C_5+ reformate obtained with a platinum–rhenium catalyst is usually higher, typically about 1 vol% higher for wide boiling range naphthas. Under certain conditions with light naphtha fractions, this result is reversed. The C_5+ yield can then be as much as 1 vol% higher for the platinum–iridium catalyst.

The attractive features of platinum–rhenium and platinum–iridium catalysts can be combined in a reforming operation. The data for the reactions of selected hydrocarbons considered earlier for platinum–rhenium and platinum-iridium catalysts indicate that the former catalyst is more selective for the conversion of cycloalkanes to aromatics, while the latter is more selective for the dehydrocyclization of alkanes. Since cycloalkane conversion occurs primarily in the initial part of a reforming system while dehydrocyclization is the predominant reaction after the cycloalkanes have reacted, it is reasonable to use a platinum–rhenium catalyst in the front of the system and to follow it with a platinum–iridium catalyst *(32).*

In such a combined catalyst operation, the high activity characteristic of a platinum–iridium catalyst can be obtained. At the same time, the C_5+ reformate yield will be more nearly equivalent to that of a platinum–rhenium catalyst in those cases where the latter has a yield advantage over a platinum–iridium catalyst, that is, in the reforming of wide boiling range naphthas.

Reforming data on a 65–190°C boiling range Persian Gulf naphtha similar to the 70–190°C fraction considered earlier are given in Figures 5.9 and 5.10 for the same platinum–rhenium and platinum–iridium catalysts used in obtaining the data in Figures 5.5 to 5.8. Data are also shown for a combined catalyst system in which a zone of platinum–rhenium catalyst is followed by a zone of platinum–iridium catalyst *(33).* In the combined catalyst system, the platinum–iridium catalyst constitutes 65% of the total catalyst charge. The runs were conducted at 28.2 atm and 488–500°C to produce 98 research octane number reformate. The naphtha had a density of 0.7421 g/cm^3 and contained (on a liquid volume percentage basis) 66.9% alkanes, 21.9% cycloalkanes, and 11.1% aromatic hydrocarbons.

Figure 5.9 shows again that the platinum–iridium catalyst is approximately twice as active as the platinum–rhenium catalyst. The activity of the combined catalyst system is intermediate between the activities of the individual platinum–iridium and platinum–rhenium catalysts.

Figure 5.10 shows that C_5+ yields are equivalent for the platinum–rhenium catalyst and the combined catalyst system and about 1.0 to 1.5 vol% higher than for the platinum–iridium catalyst. Methane and ethane yields for the combined catalyst system are higher than those for the platinum–rhenium catalyst but lower than those for the platinum–iridium catalyst. Yields of H_2 are about equivalent for the combined catalyst system and the platinum–iridium catalyst and are lower than those for the platinum–rhenium catalyst. Similarly, the yields of C_3 and C_4 hydrocarbons are about equivalent for the platinum–iridium catalyst and the combined catalyst system but are lower than the yields for the platinum–rhenium catalyst.

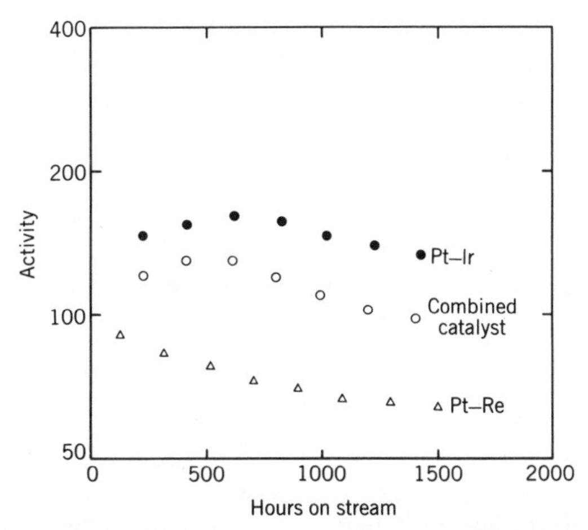

Figure 5.9 Comparison of activities of alumina-supported platinum–iridium and platinum–rhenium catalysts with that of a combined catalyst system in which platinum–rhenium is followed by platinum–iridium, the latter constituting 65 wt% of the total catalyst charge *(33)*. (Data are for the reforming of a 65–190°C boiling range Persian Gulf naphtha to produce 98 research octane number reformate at conditions of 488–500°C and 28.2 atm.) (Reprinted with permission from Elsevier Scientific Publishing Company.)

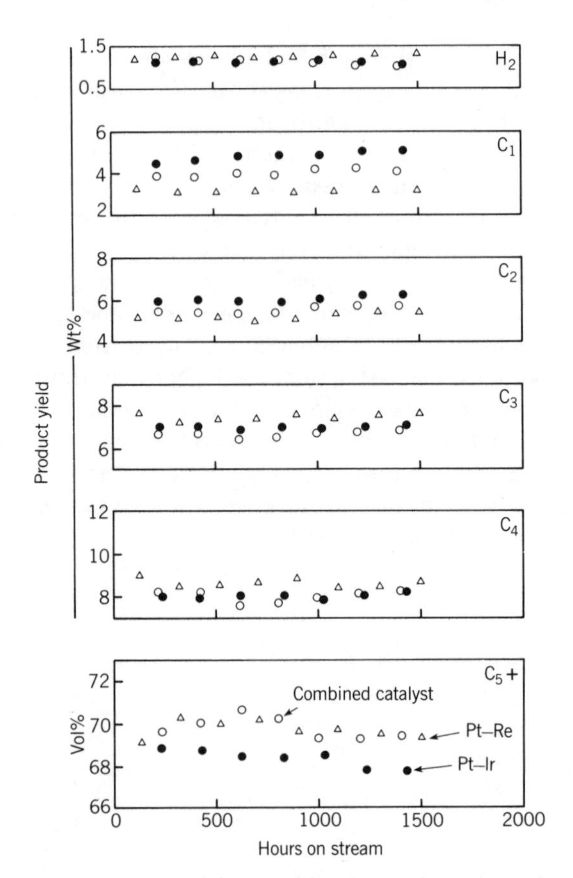

Figure 5.10 Comparison of product yields obtained in the reforming runs for which activity data are shown in Figure 5.9 *(33)*. (Platinum–iridium catalyst, •; platinum–rhenium catalyst, △; combined catalyst system, ○) (Reprinted with permission from Elsevier Scientific Publishing Company.)

The concentration of H_2 in the recycle gas was higher in the run on the platinum–rhenium catalyst than in the other two runs. Consequently, the hydrogen partial pressure at the reactor inlet was higher, averaging about 16.5 atm during the run compared to 14.2 and 14.4 atm, respectively, for the runs on the platinum–iridium catalyst and the combined catalyst system.

This comparison of the combined catalyst system with either catalyst alone is only one example. In this particular comparison, the C_5+ reformate yield obtained with the combined system was equivalent to that obtained with the platinum–rhenium catalyst alone. In other comparisons with the same distribution of platinum–rhenium and platinum–iridium catalysts in the combined system, the C_5+ yield for the combined system has been observed to be slightly lower than that obtained with the platinum–rhenium catalyst alone.

In the usual situation where the C_5+ yield is greater for the platinum–rhenium catalyst than for the platinum–iridium catalyst, the debit in C_5+ yield for the latter is substantially decreased when part of the catalyst at the front of the system is replaced by platinum–rhenium catalyst. In general, comparison of either catalyst alone with a combined catalyst system will depend on the naphtha being reformed, the reforming operating conditions, and also on the relative amounts of the two catalysts employed in the combined system.

In our discussion up to this point, comparisons of catalysts have been made with regard to activities and yields of various products. The products have included H_2, C_1–C_4 hydrocarbons, and the C_5+ liquid reformate. The yields of specific aromatic hydrocarbons in the reformate are also of interest. Data on yields of benzene, toluene, and total aromatic hydrocarbons in the reformate are summarized in Table 5.4 for platinum–iridium and platinum–rhenium catalysts in the reforming of the various naphtha fractions *(33)*. For one of the naphthas, data are also shown for the combined catalyst system. The data shown are average values for the runs.

In comparing the platinum–iridium and platinum–rhenium catalysts, we see that the yields of benzene, toluene, and total aromatic hydrocarbons in the C_5+ reformate are consistently higher for the platinum–iridium catalyst. For the combined catalyst system, the yields of benzene and toluene are equivalent to those observed with the platinum–iridium catalyst.

5.2.3 Further Remarks on Reforming Properties of Bimetallic Catalysts *(33)*

Platinum–iridium and platinum–rhenium catalysts both exhibit lower rates of deactivation in reforming than do platinum catalysts. The improved activity maintenance of platinum–iridium catalysts can be attributed, at least in part, to the lower rate of formation of carbonaceous residues on the catalyst. The decreased formation of such residues may be attributed to the increased hydrogenolysis activity resulting from the presence of iridium. It is known that the hydrogenolysis activity of iridium is much higher than that of platinum *(26)*.

The lower rate of formation of carbonaceous residues relative to that observed with a catalyst containing only the platinum is an indication of the interaction between platinum and iridium in the catalyst. It seems likely that

Table 5.4 Summary of Aromatic Hydrocarbon Yields (wt%) in Reforming with Pt–Ir and Pt–Re Catalysts *(33)*

	Catalyst		
	Pt–Ir[a]	Pt–Re[b]	Combined Catalyst[c]
From 70 to 190°C Persian Gulf naphtha[d]			
Benzene	2.6	2.1	
Toluene	13.0	11.2	
Total aromatics	46.9	46.1	
From 65 to 150°C Persian Gulf naphtha[e]			
Benzene	5.2	4.4	
Toluene	18.7	15.5	
Total aromatics	46.2	44.3	
From 65 to 190°C Persian Gulf naphtha[f]			
Benzene	4.1	3.3	4.1
Toluene	14.4	12.3	14.4
Total aromatics	50.2	50.0	51.7

NOTE: Reprinted with permission from Elsevier Scientific Publishing Company.

[a]Catalyst contains 0.3 wt% each of platinum and iridium on alumina.

[b]Catalyst contains 0.3 wt% each of platinum and rhenium on alumina.

[c]Pt–Re catalyst followed by Pt–Ir catalyst, the latter constituting 65 wt% of the total.

[d]Reforming at 28.2 atm and 490°C to produce 98 research octane number reformate. The yields are averages for 1600 hours on stream.

[e]Reforming at 28.2 atm and 488°C to produce 96 research octane number reformate. The yields are averages for 2100 hours on stream.

[f]Reforming at 28.2 atm and temperatures of 488–500°C to produce 98 research octane number reformate. The yields are averages for 1500 hours on stream.

hydrogenolysis provides a mechanism for removing the residues or their precursors from the surface.

When considering the beneficial effect of the high hydrogenolysis activity of iridium in suppressing formation of carbonaceous residues, it must be remembered that such activity may simultaneously have an adverse effect on the product distribution in reforming. A catalyst containing iridium alone on alumina is limited in this manner. For platinum–iridium catalysts, the interaction between the platinum and iridium moderates the hydrogenolysis activity of the latter. By maintaining the proper level of hydrogenolysis activity in this way, one can obtain a satisfactory product distribution and at the same time realize large advantages in activity and activity maintenance over catalysts containing platinum alone on alumina.

The lower rate of deactivation of platinum–rhenium catalysts relative to platinum catalysts cannot be attributed to a lower rate of accumulation of carbonaceous residues on the surface. For a given time on stream, the amount of such residues on the surface is not decreased by the presence of the rhenium. This point is interesting because metallic rhenium, like metallic iridium, has much higher hydrogenolysis activity than platinum *(26).* It is possible that the difference between platinum–rhenium and platinum–iridium catalysts is due to the strong retention of sulfur by the former. The inhibiting effect of sulfur on hydrogenolysis activity is well known. The improved activity maintenance of a platinum–rhenium catalyst relative to a platinum catalyst is due to better tolerance of the carbonaceous residues.

REFERENCES

1. Sinfelt, J.H. "Bifunctional catalysis." *Advan. Chem. Eng.* **5**: 37–74; 1964.
2. Sinfelt, J.H. "Catalytic reforming of hydrocarbons." In: Anderson, John R. and Boudart, Michel, eds. *Catalysis-Science and Technology.* Vol. 1. Berlin, Heidelberg: Springer-Verlag; 1981; p. 257–300.
3. Haensel, V., inventor; Universal Oil Products Company, assignee. "Alumina–platinum–halogen catalyst and preparation thereof." U.S. Patent 2,479,109. 7 pages. August 16, 1949.
4. Haensel, V., inventor; Universal Oil Products Company, assignee. "Process of reforming a gasoline with an alumina–platinum–halogen catalyst." U.S. Patent 2,479,110. 8 pages. August 16, 1949.
5. Haensel, V. and Donaldson, G.R. "Platforming of pure hydrocarbons." *Ind. Eng. Chem.* **43**: 2102–2104; 1951.
6. Jacobson, R.L.; Kluksdahl, H.E.; McCoy, C.S.; and Davis, R.W. "Platinum–rhenium catalysts: a major new catalytic reforming development." *Proceedings of the American Petroleum Institute, Division of Refining.* **49**: 504–521; 1969.
7. Sinfelt, J.H., inventor; Exxon Research and Engineering Company, assignee. "Polymetallic cluster compositions useful as hydrocarbon conversion catalysts." U.S. Patent 3,953,368. 16 pages. 1976.
8. Sinfelt, J.H. and Rohrer, J.C. "Kinetics of the catalytic isomerization–dehydroisomerization of methylcyclopentane." *J. Phys. Chem.* **65**: 978–981; 1961.

9. Rohrer, J.C. and Sinfelt, J.H. "Interaction of hydrocarbons with Pt–Al_2O_3 in the presence of hydrogen and helium." *J. Phys. Chem.* **66**: 1193–1194; 1962.
10. Sinfelt, J.H. "A simple experimental method for catalytic kinetic studies." *Chem. Eng. Sci.* **23**: 1181–1184; 1968.
11. Ciapetta, F.G.; Dobres, R.M.; and Baker, R.W. "Catalytic reforming of pure hydrocarbons and petroleum naphthas." In: Emmett, P.H., ed. *Catalysis.* Vol. 6. New York: Reinhold; 1958; p. 495–692.
12. Ciapetta, F.G. and Hunter, J.B. "Isomerization of saturated hydrocarbons in presence of hydrogenation–cracking catalysts." *Ind. Eng. Chem.* **45**: 147–155; 1953.
13. Ciapetta, F.G. and Hunter, J.B. "Isomerization of saturated hydrocarbons. Normal pentane, isohexanes, heptanes, and octanes." *Ind. Eng. Chem.* **45**: 155–159; 1953.
14. Heinemann, H.; Mills, G.A.; Hattman, J.B.; and Kirsch, F.W. "Houdriforming reactions. Studies with pure hydrocarbons." *Ind. Eng. Chem.* **45**: 130–134; 1953.
15. Hettinger, W.P.; Keith, C.D.; Gring, J.L.; and Teter, J.W. "Hydroforming reactions. Effect of certain catalyst properties and poisons." *Ind. Eng. Chem.* **47**: 719–730; 1955.
16. Mills, G.A.; Heinemann, H.; Milliken, T.H.; and Oblad, A.G. "Houdriforming reactions. Catalytic mechanism." *Ind. Eng. Chem.* **45**: 134–137; 1953.
17. Weisz, P.B. and Swegler, E.W. "Stepwise reaction on separate catalytic centers: isomerization of saturated hydrocarbons." *Science.* **126**: 31–32; 1957.
18. Hindin, S.G.; Weller, S.W.; and Mills, G.A. "Mechanically mixed dual function catalysts." *J. Phys. Chem.* **62**: 244–245; 1958.
19. Sinfelt, J.H.; Hurwitz, H.; and Rohrer, J.C. "Kinetics of *n*-pentane isomerization over Pt–Al_2O_3 catalyst." *J. Phys. Chem.* **64**: 892–894; 1960.
20. Anderson, J.R. and Baker, B.G. "The hydrocracking of saturated hydrocarbons over evaporated metal films." *Proc. Roy. Soc.* (*London*). **A271**: 402–423; 1963.
21. Carter, J.L.; Cusumano, J.A.; and Sinfelt, J.H. "Hydrogenolysis of *n*-heptane over unsupported metals." *J. Catal.* **20**: 223–229; 1971.
22. Boudart, M. and Ptak, L.D. "Reactions of neopentane on transition metals." *J. Catal.* **16**: 90–96; 1970.
23. Matsumoto, H.; Saito, Y.; and Yoneda, J. "Contrast between nickel and platinum catalysts in hydrogenolysis of saturated hydrocarbons." *J. Catal.* **19**: 101–112; 1970.
24. Anderson, J.R. and Avery, N.R. "The isomerization of aliphatic hydrocarbons over evaporated films of platinum and palladium." *J. Catal.* **5**: 446–463; 1966.
25. Dautzenberg, F.M. and Platteeuw, J.C. "Isomerization and dehydrocyclization

of hexanes over monofunctional supported platinum catalysts." *J. Catal.* **19**: 41–48; 1970.

26. Sinfelt, J.H. "Specificity in catalytic hydrogenolysis by metals." *Advan. in Catal.* **23**: 91–119; 1973.
27. Sinfelt, J.H. and Rohrer, J.C. "Kinetics of ring splitting of methylcyclopentane over alumina." *J. Phys. Chem.* **65**: 2272–2273; 1961.
28. Sinfelt, J.H. and Rohrer, J.C. "Cracking of hydrocarbons over a promoted alumina catalyst." *J. Phys. Chem.* **66**: 1559–1560; 1962.
29. Coonradt, H.L. and Garwood, W.E. "Mechanism of hydrocracking." *Industrial and Engineering Chemistry, Process Design and Development.* **3***(1)*: 38–45; 1964.
30. Sinfelt, J.H. "Esso catalyst based on multimetallic clusters." *Chem. Eng. News.* **50**: 18–19; July 3, 1972.
31. Cecil, R.R.; Kmak, W.S.; Sinfelt, J.H.; and Chambers, L.W. "Developments in powerforming with advanced catalysts." *Proceedings of the American Petroleum Institute, Division of Refining.* **52**: 203–213; 1972.
32. Sinfelt, J.H., inventor; Exxon Research and Engineering Company, assignee. "Combination reforming process." U.S. Patent 3,791,961. 4 pages. 1974.
33. Carter, J.L.; McVicker, G.B.; Weissman, W.; Kmak, W.S.; and Sinfelt, J.H. "Bimetallic catalysts; application in catalytic reforming." *Appl. Catal.* **3**: 327–346; 1982.
34. Carter, J.L. and Sinfelt, J.H., inventors; Exxon Research and Engineering Company, assignee. "Startup method for a reforming process." U.S. Patent 4,220,520. 5 pages. 1980.
35. Sinfelt, J.H., inventor; Exxon Research and Engineering Company, assignee. "Serial reforming with platinum catalyst in first stage and iridium, rhodium, ruthenium or osmium catalyst in second stage." U.S. Patent 3,684,693. 3 pages. 1972.
36. Sinfelt, J.H., inventor; Exxon Research and Engineering Company, assignee. "Reforming with a single platinum group metal." U.S. Patent 3,871,996. 6 pages. 1975.

Chapter 6

CONCLUDING REMARKS

The research on bimetallic catalysts described in this monograph was conducted with the joint goals of fundamental understanding and application. New catalytic phenomena and catalyst systems were discovered, and new concepts concerning catalyst structure emerged. Our understanding of bimetallic catalysts was achieved with a combination of chemical and physical probes. In the early stages of the research on bimetallic clusters, chemical probes were especially important, since the physical probes then available were generally unsatisfactory for investigating highly dispersed catalyst systems. The results of these early studies demonstrated the great power of purely chemical probes for the investigation of catalyst structure. For example, measurements of rates of hydrogenolysis of ethane to methane provided convincing evidence for the existence of bimetallic clusters and simultaneously led to valuable insight into the structures of these entities. When a generally applicable physical probe, namely, extended x-ray absorption fine structure (EXAFS), was eventually developed for the investigation of bimetallic clusters, the results obtained with it strongly confirmed the conclusions derived from the early studies employing chemical probes. The bimetallic cluster systems provided an excellent test of the EXAFS method for investigations of catalyst structure.

The industrial application of bimetallic catalysts has been huge, especially in the petroleum industry. Their use in the catalytic reforming of petroleum fractions has been vital for the production of unleaded gasoline. The bimetallic reforming catalysts currently in use have exhibited outstanding performance in refineries throughout the world. The area is still an intriguing one for further research, and it is highly probable that improvements in the existing systems will emerge as a result of future research.

Possibilities for applications of bimetallic catalysts to reaction systems other than reforming would appear to be excellent, in view of the many possible combinations of metallic elements. The concept of bimetallic clusters has introduced much versatility into the area of metal catalysis. Moreover, it is possible to develop bimetallic catalysts with a rational, scientific basis. The feasibility of this type of endeavor has been aided greatly by the formulation of useful working concepts and by the development of powerful new methods for investigating catalytic materials.

INDEX